Strategic Alliances for Innovation and R&D

A volume in
Research in Strategic Alliances

Series Editor:
T. K. Das, *City University of New York*

Research in Strategic Alliances

T. K. Das, Series Editor

Researching Strategic Alliances: Emerging Perspectives (2010)
edited by T. K. Das

Strategic Alliances in a Globalizing World (2011)
edited by T. K. Das

Behavioral Perspectives on Strategic Alliances (2011)
edited by T. K. Das

Strategic Alliances for Value Creation (2012)
edited by T. K. Das

Management Dynamics in Strategic Alliance (2012)
edited by T. K. Das

Managing Knowledge in Strategic Alliances (2013)
edited by T. K. Das

Interpartner Dynamics in Strategic Alliances (2013)
edited by T. K. Das

Managing Public-Private Strategic Alliances (2014)
edited by T. K. Das

Strategic Alliances for Innovation and R&D (2014)
edited by T. K. Das

Strategic Alliances for Innovation and R&D

Edited by

T. K. Das
City University of New York

IAP
INFORMATION AGE PUBLISHING, INC.
Charlotte, NC • www.infoagepub.com

Library of Congress Cataloging-in-Publication Data

Strategic alliances for innovation and R & D / edited by T.K. Das.
pages cm -- (Research in strategic alliances)
Includes index.
ISBN 978-1-62396-622-5 (paperback) -- ISBN 978-1-62396-623-2 (hardcover) --
ISBN 978-1-62396-624-9 (e-book) 1. Strategic alliances (Business) 2.
Research, Industrial. 3. Technological innovations. 4. New products. I.
Das, T. K.
HD69.S8S7734 2014
658.5'7--dc23

2014002436

Printed in the United States of America

CONTENTS

FOREWORD TO THE SERIES

Relationships have been important to commercial activity and economic transactions for thousands of years. Yet, the development of a global competitive landscape has substantially enhanced the importance of partnerships between economic entities. These partnerships, referred to as strategic alliances, provide access to resources and capabilities that allow firms to gain economies of scope and to increase their productivity and innovation. The economies, productivity and innovations are necessary to at least maintain competitive parity and especially to achieve a competitive advantage in the often highly competitive global markets. Strategic alliances have also become a prominent means of entering new markets, especially foreign markets. Therefore, alliances and the networks of firms of which they are a part have become essential for the conduct of business for all types of firms, large, small, established and new.

Because of their growing importance, research on strategic alliances has increased markedly in the last 2 decades. Yet, there is need for an authoritative compendium of strategic alliance research and knowledge. This book series on *Research in Strategic Alliances* fills this critically important gap in our field. It provides a thorough examination of significant topics that provide complete and up-to-date knowledge on strategic alliances. This book series will serve as a catalyst for more effective management of strategic alliances and will guide future research on them. I commend it to you.

—Michael A. Hitt
Distinguished Professor and Joe B. Foster Chair in Business Leadership at Texas A&M University, and Past President of the Academy of Management and the Strategic Management Society.

Strategic Alliances for Innovation and R&D, pp. vii–vii

ABOUT THE SERIES

The globalization of markets has led to increased interdependence among business firms, leading to an explosion in the number of strategic alliances. Strategic alliances, briefly, are cooperative arrangements aimed at achieving the strategic objectives of two or more partner firms. These interfirm arrangements can range from joint R&D to equity-based joint ventures. However, the scholarship relating to strategic alliances remains largely dispersed in the literatures of traditional academic disciplines such as strategic management, marketing, economics, and sociology. This book series on strategic alliances will cover the essential progress made thus far in the literature and elaborate upon fruitful streams of scholarship. More importantly, the book series will focus on providing a robust and comprehensive forum for new scholarship in the field of strategic alliances. In particular, the books in the series will cover new views of interdisciplinary theoretical frameworks and models (dealing with resources, risk, trust, control, cooperation, learning, opportunism, governance, developmental stages, performance, etc.), significant practical problems of alliance organization and management (such as alliance capability, interpartner conflict, internal tensions, use of information technology), and emerging areas of inquiry. The series will also include comprehensive empirical studies of selected segments of business, economic, industrial, government, and non-profit activities with wide prevalence of strategic alliances. Through the ongoing release of focused topical titles, this book series will seek to disseminate theoretical insights and practical management information that will enable interested professionals to gain a rigorous and comprehensive understanding of the field of strategic alliances.

—T. K. Das
City University of New York
Series Editor, *Research in Strategic Alliances*

Strategic Alliances for Innovation and R&D, pp. ix–ix

CHAPTER 1

THE GOVERNANCE OF HIGH-TECH ALLIANCES

Trust, Contracts, Cultural Differences, and Interdependence

Gjalt de Jong, Nora Balogh, and Rosalinde Klein Woolthuis

ABSTRACT

Prior work has established the importance of trust and contracts for understanding the performance of strategic alliances. Despite all these efforts, however, the need for formal and informal governance mechanisms is still the subject of ongoing debate due to inconclusive findings. We suggest that the alliance literature has largely overlooked organizational cultural differences and interdependence as important determinants of trust and contracts. We will argue that the impact of cultural differences and interdependence is different for trust and contracts. This difference may help to explain some of the inconclusive findings in the alliance literature and may also help to assess the trade-off between trust and contracts in strategic alliances. Our study is among the first to examine the role of cultural differences and interdependence in high-tech alliance management, using a unique dataset

Strategic Alliances for Innovation and R&D, pp. 1–27

from 391 Dutch alliances. The results provide convincing support for our approach to the study of high-tech alliance management.

INTRODUCTION

Strategic alliances offer companies opportunities to survive competition in ever-increasingly globalizing markets (Das & Kumar, 2011a; Das & Teng, 2001). Previous research has established that alliance management contributes to the success of interfirm collaboration (Ariño & Reuer, 2004; Lowensberg, 2010). Alliance management is not a product that can be bought on the market, which implies that collaborating firms need to design and implement governance mechanisms themselves (Artz & Brush, 2000; de Jong & Klein Woolthuis, 2009). According to the alliance literature, there are two main governance mechanisms that firms can develop and use, each corresponding to considerable areas of research (Das & Kumar, 1998, 2007; Das & Teng (1998). One line of research indicates a need for formal governance (i.e., contracts) for the management of interfirm collaboration (de Jong & Klein Woolthuis, 2010). Another line of research highlights the importance of relational governance (i.e., trust) for successful alliances (Brockner, Siegel, Daly, Tyler, & Martin, 1997; Chen, 2000). Despite all these efforts, however, the management of strategic alliances is still subject to ongoing debate as a result of mixed empirical findings, among other reasons, (Contractor, 2005; Das & Kumar, 2011b), which thus calls for more studies that investigate antecedents of and relationships between formal and informal governance mechanisms.

To some extent, differences in concepts, measurements, methods, and samples can explain the inconclusive empirical findings (de Jong & Nooteboom, 2000). These explanations notwithstanding, we suggest that alliance governance research suffers from a missing variable bias. In other words, existing research models lack variables to comprehend how alliance partners apply formal and informal mechanisms for the governance of the interfirm collaboration. Our first contribution to the literature is to propose organizational cultural differences and interdependence as being the key missing variables in alliance governance research. More especially, we argue that cultural differences and interdependence are inherent in any alliance but particularly in high-tech collaboration, which is the focus of this research. We argue that cultural differences and interdependence have a different impact on trust and contracts: this may serve as an explanation for some of the mixed empirical findings. Although cultural differences and interdependence have been acknowledged as important factors in international business (e.g., Beugelsdijk, Koen, & Noorderhaven, 2009) and in strategic collaboration (e.g., Das & Kumar, 2007), whether and how they determine formal and informal governance mechanisms is not, at least not explicitly, addressed in the alliance literature. We aim to fill this research gap.

Our second contribution concerns the empirical study. Notable exceptions notwithstanding, large-scale empirical studies of trust and contracts, in general, and in combination with interdependence and cultural differences, in particular, are few and far between. Among other reasons, this is due to the confidentiality involved in business contracts, which hampers data collection. Our empirical study focuses on business relationships between firms that operate in high-tech industries (biotechnology, new material development, information technology, maritime technologies, and environmental technology). In these industries, strategic alliances are almost a prerequisite for innovation and successful firm performance. The life cycle of R&D in these industries is usually short. Much of the new technological knowledge quickly becomes outdated, often even before it has been incorporated into new products and/or services. Hence, in high-tech industries in particular, we find many collaborative efforts between firms, including rival firms. Our research setting is also relevant because high-tech alliances are almost inherently characterized by interdependence and organizational cultural differences.

The outline of this chapter is as follows. In the next section, we will explain the theoretical foundations of our study. A discussion of the research methodology and our empirical results will then follow. We will conclude with an appraisal of the results and suggest avenues for future research.

THEORETICAL BACKGROUND

Alliance governance research has offered valuable insights concerning the role of contracts and trust in strategic alliances (Lyons & Mehta, 1997; Mayer, Davis, & Schoorman, 1995; Nooteboom, 2004). In what follows, we will first review the definitions of and main findings for contracts and trust, as well as the relationships between these two governance mechanisms. Subsequently, we will discuss insights related to the main characteristics of high-tech alliances, that is, cultural differences and interdependence.

Contracts and Trust

Strategic alliances "include voluntarily collaborating firms who engage in exchange, sharing, or co-development, and include contributions by partners of capital, technology, or firm-specific assets" (Gulati & Singh, 1998). In principle, all strategic alliances can be thought of as coalignments between two or more firms (Lei, Sloum, & Pitts, 1997); the interests are aligned to achieve a common goal. This is not without risks; firms are especially subject to the risk of opportunism. Opportunism refers to incomplete or distorted disclosure of information, and especially to calculated efforts to mislead the partner (Hill, 1990; Williamson, 1985). Transaction cost economics states that the threat of opportunism may be curbed in social exchanges by designing a detailed formal contract between the parties (Gulati, 1995; Poppo & Zenger, 2002). This legal document provides a framework, within which cooperation between partners proceeds by specifying the mutual rights and obligations of the partners, as well as a specification of their input and expected output, the processes involved in their exchange, and dispute-resolution mechanisms (Reuer & Ariño, 2007).

Hence, the design of a contract is important in order to mitigate *ex ante* and *ex post* transaction costs (de Jong & Klein Woolthuis, 2008a). Contract research indicates a need to develop an appropriate measurement for the content and the role of contracts in strategic alliances. Alliance research has shown that the level of detail in alliance contracts varies greatly and that contracts may have different roles or functions. Transaction cost economics argues that the level of contractual detail depends on the extent to which the alliance partners face opportunism; more contractual safeguards are needed when the risks of opportunistic behavior are great. Research has shown that contracts are indeed used to safeguard positions, as suggested by transaction cost economics, but also to coordinate activities or to mitigate external contingencies (de Jong & Klein Woolthuis, 2009; Mellewigt, Madhok, & Weibel, 2007).

Economic theory argues that the risk of opportunism stems from the inherent untrustworthiness of exchange partners. That is, the concept of "trust" among alliance partners is in fact a manifestation of their pure self-interest to behave in a trustworthy manner, since the alliance contract serves as an incentive for them to behave in this way (Barney & Hansen, 1994). Alliance research has shown that this perspective in terms of trust is too narrow. According to a more behavioral perspective, transactions are embedded in social relationships (Baker, Gibbons, & Murphy, 2002). Trust represents one partner's expectation that the other will not exploit their vulnerabilities, even when presented with the opportunity to do so (Krishnan & Noorderhaven, 2006).

Several scholars have argued that trust is a multidimensional construct with multiple antecedents (Aulakh, Kotabe, & Sahay, 1996; de Jong & Klein Woolthuis, 2008b; Mayer et al., 1995; Nooteboom, 2002). These studies highlight, for example, the importance of open information sharing and flexibility. Because trust develops in the course of the process of collaboration, it is likely to be different at the various stages of interfirm collaboration (Ariño, De la Torre, & Ring, 2001). Nooteboom (1996), for instance, refers to trust in people or organizations as behavioral trust, and differentiates among trust in competences (competence trust), trust in intentions, and honesty or truthfulness. Trust may have different antecedents such as continuity expectations, flexibility, information exchange, and control mechanisms. Although trust primarily exists between individuals, alliance governance research has shown that it can be applied to organizations as well (Aulakh et al., 1996).

It is acknowledged that trust and contracts coexist and interact with each other in the context of strategic alliances (Baker et al., 2002; Contractor, 2005; Lyons & Mehta, 1997), albeit the relationship between the constructs is subject to debate. One line of research suggests that contracts and trust are substitutes for each other in alliance governance (Dyer & Chu, 2003; Gulati, 1995; Gulati & Singh, 1998; Reuer & Ariño, 2007). When trust acts as a means of curbing transaction costs, it reduces the need to write a detailed contract, since a large proportion of future contingencies can be resolved on the basis of trust. In such cases, trust substitutes for detailed contracts. The arguments for complementarity in terms of trust and contracts stem from the key characteristic of contracts, that is, that contracts are inherently incomplete. As it is impossible to foresee all future contingencies and incorporate them into the contract, alliance partners have to rely on trust. Hence, trust becomes important where contracts no longer suffice (Nooteboom, 2004).

The aforementioned studies differ in the sign of the relationship between trust and contracts that is hypothesized. This derives from contradictory assumptions about the causality of the governance mechanisms. It is a question of whether trust determines contracts, or conversely, whether the characteristics of the contract are what shape trust. Klein Woolthuis, Hillebrand, and Nooteboom (2005) show that, although theoretical papers assume that the legal framework predicts the level of trust, empirical findings generally report the opposite. Based on in-depth case studies, Klein Woolthuis et al. argue that in case trust and contracts are complementary governance tools, the contract is the basis for trust to develop. In a similar vein, Nooteboom (2002) also argues that contracts are one of the sources of reliability in exchanges, with reliability referring to trustworthiness in the broadest sense.

That said, large-scale empirical studies of trust and contracts are few and far between, leaving the question about the relationship between trust and contracts largely unresolved. Our study will serve to complement prior alliance governance research as follows. We aim to reveal how trust and contracts are affected by interdependence and cultural differences. To do this, we designed two models (that is, one for each of the two governance forms), and studied whether or not the relationship between the key characteristics of high-tech alliances and each of the governance forms is different. Although interdependence and cultural differences are identified as important features of international business in general (Beugelsdijk, Koen, & Noorderhaven, 2009; Denison & Mishra, 2005) and have been used to explain alliance performance differentials (Guidice & Cullen, 2007; Sambasivan, Loke, Mohamed, & Leong, 2011), they have not been related to alliance governance in terms of trust or contracts, which is the key focus of research in this study.

Organizational Culture and Interdependence

Beugelsdijk et al. (2009) define organizational culture as "the shared values and beliefs that help individuals understand organizational functioning ... and help managers direct the course of the organization more effectively" (p. 313). For two reasons, this definition is helpful for our research. First, since organizational culture is based on a sentiment similar to trust, it is likely to influence the behavior and beliefs of individuals. Second, although it is a sentiment, it can be applied as a managerial tool to develop the organization; in other words, it can be applied as a coordination device, such as the alliance contract. From this perspective, the organizational culture of a focal company is likely to determine the governance mechanisms of the alliance, in which they participate.

Various scholars have argued that organizational culture may determine alliance performance directly (Beugelsdijk et al., 2009; Pothukuchi, Damanpoor, Choi, Chen, & Park, 2002). Yang and Lin (2005), for example, find that the higher the degree of strategic and technical distances between firms, the higher the probability that firms will choose an alliance instead of a merger. In other words, organizational culture is potentially important, because the management styles of the partnering companies are reflected in the management of the alliance. If, for example, cultural distance creates ambiguous expectations about the behavior of the partner firm, trust might be difficult to develop or to maintain. Furthermore, a lack of similarity in the organizational cultures of the partnering firms may induce collaborating firms to request more guarantees to be specified in a contract.

According to the resource-based view, interorganizational relationships enable firms to attain a mutually beneficial outcome that they could not have reached on their own (Lowensberg, 2010; Mellewigt et al., 2007). If organizations depend on the resources of their exchange partners in achieving their goals, then such strategic alliances entail dependence on the part of the collaborating firms. Hence, dependence refers to a firm's need to maintain an exchange relationship with other firms in order to achieve its desired goals (Razzaque & Boon, 2008). Interdependence refers to the belief that the outcome of a relationship depends on the separate as well as joint actions of the members of the cooperative project. Like trust, interdependence is a multidimensional construct. Sambasivan et al. (2011), for example, differentiate between the extent to which firms have to rely on their partners to complete their tasks (task interdependence) or to reach their goals (goal interdependence) in the alliance, and how their individual success influences the performance of the collaboration (reward interdependence). One important aspect of mutual dependence is that it entails a firm's exposure to changes in its partner's behavior and actions. Since this exposure may lead to perceived vulnerability in a cooperative relationship, the level of trust or contractual detail is likely to be affected by the level of interdependence among the partner firms.

HYPOTHESES

Economic models of transactions stem from the generic idea that organizations emerge to curb transaction costs (Gulati, 1995). Firms will formally organize their transactions with other companies in order to monitor the hazards associated with exchange, such as opportunism. Williamson (1985) focuses on the institutional arrangements of interfirm exchange or, more specifically, formal contracts. Whereas Williamson denies the importance of trust in business relationships, other scholars have argued and shown that trust does have value and that it is useful in strategic alliances (Nooteboom, 2002). It goes without saying that alliances have risks that need to be mitigated, for which contracts are applied. At the same time, collaborative relationships are typically associated with trust between alliance partners. We will therefore use both perspectives for the development of our hypotheses. Figure 1.1 presents our research model which consists of two parts: one for interorganizational trust and one for contractual complexity.

Model I. Interorganizational Trust

Model I is concerned with interorganizational trust, defined as the expectation that a partner will not engage in opportunistic behavior, even in the face of opportunities and incentives for opportunism, irrespective of the ability to monitor or control that party (Krishnan & Noorderhaven,

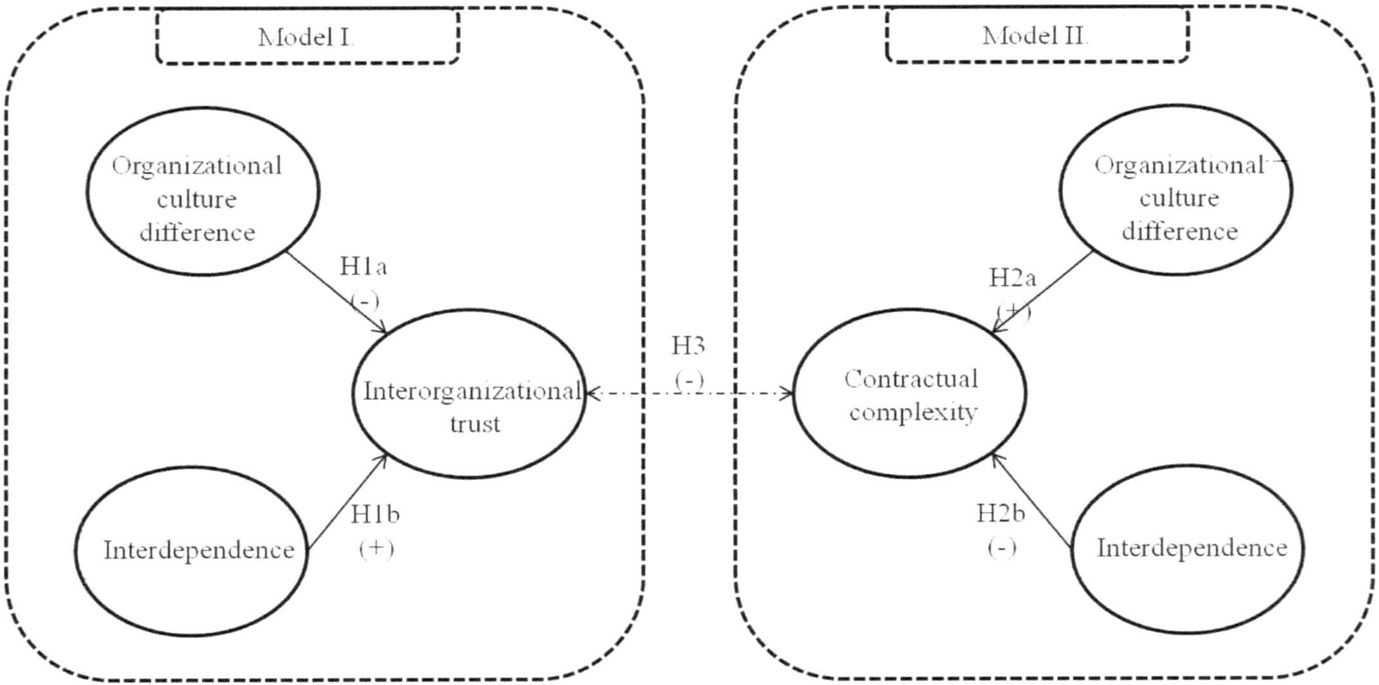

Figure 1.1. The research model of this study (with expected signs in parentheses).

2006). The model examines how cultural differences and interdependence affect interorganizational trust.

Organizational culture has been defined as the norms and beliefs that bind organizations together. These values help managers to communicate goals and strategy throughout the organization (Marcoulides & Heck, 1993). It is argued that when companies engage in collaboration, the outcome of their cooperative project and the satisfaction associated with their partnership depend on the similarities in their organizational culture (Beugelsdijk et al., 2009; Pothukuchi et al., 2002). When alliance partners encounter organizational cultural differences, this may lead to misunderstandings and interaction problems. Gulati and Sytch (2008), for example, argue that organizational similarity makes partners more attractive in the partner selection process. Similarities in organizational practices foster knowledge-driven trust stemming from repeated interaction. Partners with dissimilar organizational cultures need time to develop common managerial practices and may therefore experience lower trust than culturally similar partners (Potchuchi et al., 2002). Organizational differences in culture may also be associated with lower commitment (Veiga, Lubatkin, Calori, & Very, 2000). Building trust might be more difficult when organizational cultures are highly dissimilar, since homogenous expectations and shared assumptions about the alliance may not readily exist (Parkhe 1998a, 1998b). These arguments lead to:

> **Hypothesis 1a:** *Organizational cultural differences have a negative effect on interorganizational trust.*

Alliance research has shown that in the presence of trust, parties can cooperate with each other more fluently, as trust lowers the transaction costs inherent in exchanges (Aulakh et al., 1996; Krishnan & Noorderhaven, 2006). Trust can be understood as the willingness to rely on the behavior of others and the willingness of a party to be vulnerable to the actions of another party, based on the assumption that the other party will perform a particular action important to the trustor (Mayer et al., 1995). This definition of trust implies that a trusting relationship is associated with the acceptance of the vulnerability that stems from the dependence on the partner. In a similar vein, Razzaque and Boon (2008) argue that a trusting relationship makes the dependence inherent in the alliance more acceptable to partners, since they then feel assured that the desired outcome of the partnership will be achieved. Moreover, a symmetric high-dependence relationship engenders valuable resources as well as cooperation between partners. Razzaque and Boon therefore conclude that when mutual dependence in the partnership is high, satisfaction with the alliance is more likely to be achieved in the presence of trust. Smith and Barkley (1999) argue that greater interdependence between partners leads to greater mutual trust in their working relationship. As partners to an alliance depend highly on their partners to achieve their goals, they experience high exit barriers, which is a strong motivation for them to build a strong relationship. Trust and commitment emerge when the interdependence structure between the partners leads to a convergence of the interests of the participants (Kumar, Scheer, & Steenkamp, 1995). Ariño et al. (2001) also define interdependence as a key element of interorganizational trust and suggest that trust develops in the presence of the vulnerability resulting from the dependence inherent in cooperation. The arguments above indicate that trust emerges as a result of the interdependence inherent in the relationship, because partners have valuable resources to lose should they behave opportunistically; therefore, interdependence fosters trust between the partner firms. These arguments lead to:

> **Hypothesis 1b:** *Interdependence has a positive effect on interorganizational trust.*

Model II. Contractual Complexity

In Model II, the central point of the analysis is how organizational cultural differences and interdependence influence the extent of contractual complexity in strategic alliances. Contractual complexity, as it has been pointed out, is a function of the increasing risk of opportunistic behavior (Reuer & Ariño, 2007). Relational risk in alliances can be mitigated by adapting a formal contract in alliances (Mellewigt et al., 2007). Relational risk relates to whether or not cooperation between alliance partners is effective (Das & Teng, 1996).

We argue that organizational cultural differences may lead to conflicts that may create obstacles and a greater need to mitigate risks by means of a detailed contract. Organizational cultural differences might also translate into different assessments of control styles (Pothukuchi et al., 2002). If companies have a different understanding about management monitoring and control, a specific and detailed contract that specifies alliance responsibilities and roles counterbalances inconsistencies in management styles. Marcoulides and Heck (1993), for example, identify organizational culture as a set of sociocultural systems that define an organization. One of these systems refers to the formal organizational structure of a company, which can also be understood as managerial practices including those associated with contracting. Hence, organizational cultural differences will lead to conflicts and a need to bridge different management styles, which are reflected in a more detailed contract in order to mitigate risks, define roles and responsibilities, and reduce contingencies. These arguments lead to:

Hypothesis 2a: *Organizational cultural differences have a positive effect on contractual complexity.*

Interdependence is inherent in strategic alliances as firms need to rely on their partners to achieve their goals. The extent to which a firm is dependent on its partner and the alliance project may differ. We hypothesize that variations in dependence explain variations in contract complexity. Guidice and Cullen (2007) examine the relationship between interdependence, control, and alliance performance. They find that the use of formal contracts will reduce performance in alliances characterized by high interdependence, whereas the use of informal controls will improve performance. Guidice and Cullen argue that if mutual dependence in a strategic alliance is high, the partners in the cooperative project have much to lose from opportunistic behavior, since they rely heavily on their partner and the alliance projects to achieve their firm-specific and collaborative goals. Hence, mutual dependence acts as a control mechanism against opportunism, since the possibility of losing a valuable and resourceful partnership acts as a self-enforcing sanction (Reuer & Ariño, 2007). In the presence of interdependence, complex contracts also become less attractive, because a detailed contract entails high *ex post* transaction costs due to high monitoring and/or contract renegotiation costs. High interdependence makes it extremely dangerous for the partners to behave opportunistically (Kumar et al., 1995); it therefore minimizes the risk of a firm's acting in its self-interest, since any move which adversely affects a partner would adversely affect the alliance as a whole (Sambasivan et al., 2011). These arguments show that interdependence may act as a means to limit opportunistic behavior and therefore reduce the need to specify complex contracts. These arguments lead to:

Hypothesis 2b: *Interdependence has a negative effect on contractual complexity.*

Trust and Contracts as Substituting Governance Mechanisms

The alliance literature offers different perspectives on the relationship between trust and contracts, accompanied by mixed empirical findings.

The nature of the trust-contract relationship is subject to an ongoing debate with conflicting views about whether there is a substitution or complementary relationship between the two governance forms. We will contribute to this debate as follows. We argue that organizational cultural differences will lead to conflicts that can be solved by developing a complex contract (Hypothesis 2a), but at the same time we also hypothesize that this will decrease interorganizational trust (Hypothesis 1a). Hence, given that organizational cultural differences will decrease interorganizational trust, a detailed formal contract can substitute for trust as a governance mechanism in the case of major cultural differences. A similar line of reasoning applies to the hypotheses concerning interdependence. We hypothesize that interdependence will foster trust (Hypothesis 1b), but at the same time it will also decrease the need to specify a detailed contract (Hypothesis 2b). Thus, given that interdependence reduces the need for a detailed contract, trust can substitute for contracts in the case of high interdependence. Our final hypothesis is based on the assumption that the hypotheses of Models I and II are accepted, leading to:

Hypothesis 3: *Contracts and trust act as substitute governance mechanisms.*

METHODS

Data Collection and Sample

Our data collection proceeded in two phases (see also de Jong & Klein Woolthuis, 2008a, 2008b). In the preparatory phase of the fieldwork, we conducted 25 semistructured interviews with consultants of the Dutch Ministry of Economic Affairs, who were involved in policy programs to stimulate interfirm collaboration in terms of innovation. Additionally, the consultants selected 20 cases (10 successful and 10 less successful ones) that we studied in great detail to obtain in-depth knowledge of the high-tech collaboration. Case research is suitable for exploratory research where understanding is the primary objective. The 20 cases dealt with collaborative innovation and hence involved complex transactions, for which close collaboration between partners was necessary over a considerable period of time. The cases involved legally independent partners who shared costs and benefits more or less evenly. All cases entailed uncertainty and/or complexity, and specific assets, and hence risks of dependence, opportunism, and "hold-up." Under strict confidentiality, we received full access to all documents in the cases—including not only the interfirm contracts but also project plans, annual reports of the companies involved, personal notes and letters, and half-yearly progress reports—that were available at the Ministry. Among other things, this allowed us to examine the contracts with respect to the exact content of each clause laid down in them. In addition, clippings from newspapers and trade magazines concerning the collaborations were collected. To enable comparison between the cases and to ensure the quality of the case analysis, a case protocol was written in order to describe the alliance's history, development, and outcome. The interviews with the consultants were transcribed into interview reports and sent back for verification and agreement. Hence, all this allowed us to reconstruct the development of high-tech alliances and to check the data from the interviews with the secondary sources. We used this information to design our survey. The survey was field-tested using a sample of ten companies involved in R&D alliances. This resulted in a number of modifications to the questionnaire.

In the second stage, a research team conducted telephone interviews with 572 business managers of interfirm R&D collaboration. Prior to these interviews, all managers received an explanatory letter inviting them to participate. We briefed the team on the features of R&D, high-tech industries, and interfirm relationships. The team made three attempts to identify and interview the selected respondents. The case firms were identified from a database of Dutch interfirm high-technology alliances published by the Ministry of Economics Affairs. This enabled us to identify the business managers who were responsible for interfacing with the partner firms. They were considered to be the informants most knowledgeable about the interfirm relationships. During the interview, main topics such as the history and purpose of the alliance as well as contracts, investments, and industry dynamics were discussed. One of the first questions required the respondents to identify the business partner in the alliance in question. We used this information to cross-validate the information from the database. Because high-tech alliances are typically concerned with specific projects and goals, we also asked the respondents to identify the one project that was the most important to the interfirm alliance. By focusing on interfirm collaboration within one sector (high-tech industries), we reduced the range of extraneous variations such as the level of uncertainty or competition that might influence the constructs of interest. Some open questions were added to enable the respondents to tell their own story to some extent. In total, 50 main questions (often divided into several sub-questions) were asked. One outcome of this was that the interviews that were designed to take half an hour would sometimes take up to an entire hour, depending on the respondent.

We obtained 391 useable responses, giving an effective response rate of 68.5%. This rate is considerably higher than those observed in prior studies on interfirm relationships, which were usually in the 10 to

33% range (Dyer & Chu, 2003; Gulati & Sytch, 2008; Poppo & Zenger, 2007). It was also satisfactory considering that this study required direct senior management involvement and the confidentiality of some of the information requested.

The nonresponse is low (31.5%) especially considering that only 10.5% actually refused to be interviewed. Out of the respondents, 20.1% could not be contacted within the three attempts that the interviewers made to try to get in touch with them. To investigate whether the nonresponse incurs a bias, the noncooperating respondents (10.5%) were asked for their reasons for not participating. The reasons for refusal were, on the one hand, a lack of time and interest, and, on the other hand, irritation because they had recently taken part in another survey. Although these reasons can hide their true motive for not participating in the survey (such as an unsuccessful cooperative project), the low nonresponse and the reasons for not participating do not raise serious doubts as to the implications of the nonresponse.

Studies of this type, where one asks the same respondent about both dependent and explanatory variables, are in principle vulnerable to common method bias, in particular common source bias (Podsakoff & Organ, 1986). In terms of the aim of this study, reliance on key informants such as our respondents seems to be the only realistic and feasible way to obtain the required information (Huber & Power, 1985). We implemented the following steps to address possible validity concerns. First, we used a complex model specification with two separate models (estimated with different techniques), one that is recognized as among the most important solutions to prevent common source bias. Second, available data were able to be tested for convergence by triangulation using secondary data. We compared the outcomes of the self-reported data in the questionnaire with the archival data on the 20 cooperative projects that we studied in the first phase of the data collection. The congruence of the data from the questionnaires and case studies supported the accuracy of the reported data. Third, we asked questions concerning objective events rather than subjective expectations or opinions. Fourth, via the sequence of our questions we aimed to minimize the effects of consistency artifacts. Some suggest letting the independent variable follow, rather than precede, the independent variables. Podsakoff and Organ (1986), however, argue that correlations will be similar using either method. In our opinion, a life-cycle approach would best serve an accurate reflection of the interfirm collaboration. Hence, for the purpose of this study, we structured the questions in the survey from past interactions through partner selection, contract negotiations, contract execution, and outcomes of the interfirm collaboration. Fifth, we conducted Harman's single factor test and this gave no indication of common method bias. Exploratory factor analysis yielded 12 factors with eigenvalues greater than 1. The principal component factor resulted in 16 factors passing the Kaiser criterion of eigenvalues greater than 1. In sum, it is unlikely in our case that the findings can be attributed to common source bias.

Measurements

Table 1.1 provides an overview of the items that we used to measure the constructs of our theoretical model.

Table 1.1. Concepts and Measurements

Constructs, Items, and Scales
Interorganizational trust (alpha = 0.73)
1. During the project, our partners treated our problems constructively and with care.
2. I never had the feeling of being misled.
3. We talked openly and informally with our partner about our ideas, feelings, and interest.
4. We provided each other with all the information that was relevant to the project.
5. Criticism could be openly aired if this contributed to the completion of the project.
(1 = strongly disagree, 5 = strongly agree)
Contractual complexity (alpha = 0.83)
Please indicate whether one or more of the following agreements are present in the contract with you partner:
1. Relationship goal and outcome
2. Relationship duration
3. Project plan of the relationship
4. Investments by all alliance parties (human, material, and financial resources, and knowledge)
5. Risk allocation (internally as well as externally in terms of possible customers)
6. Project management: Which partner has project leadership, when and how do parties inform each other, how do they communicate, and how is the project monitored
7. Pledge of secrecy: protection of know-how and sanctions in case of monopolizing knowledge and/or breech of agreement
8. Ownership of the final product or technology
9. Ownership of the final method
10. License agreement concerning the exploitation of all alliance results
11. Patent rights of all alliance results
12. Relationship adjustments and/or termination arrangements under unforeseen circumstances such as disappointing market potential
13. Arrangement as to how parties will deal with any future conflicting interests

(Table continues on next page)

Table 1.1. Concepts and Measurements (Cont.)

Constructs, Items, and Scales
Interdependence ($r = 0.26, p < .05$)
1. We gave our partner access to new customers and markets, which was very important to us.
2. Our partner gave us access to new customers and markets, which was very important to them.
(1 = strongly disagree, 5 = strongly agree)
Organizational cultural differences ($r = 0.57, p < .01$)
1. Cultural differences between us and our partner have resulted in problems.
2. Due to inefficient ways of communicating, problems often occurred with our partner.
(1 = strongly disagree, 5 = strongly agree)
Firm size ($r = 0.76, p < .01$)
1. What is the number of employees in your firm? (1 = 0 –10; 3= 100-250; 5 = >1000)
2. What was your turnover in 1997? (1: < 500.000; 3 = 1 –10M; 5: > 50M)
Reputation ($r = 0.37, p < .05$)
1. Our partner had a very good name and reputation in our industry.
2. The reputation of our partner played an important role in our choice of partner.
(1 = strongly disagree, 5 = strongly agree)
Previous relationship history ($r = 0.57, p < .01$)
1. Our current relationship is a continuation of a previous, long-term relationship.
2. Before this project, a friendly relationship with the partner was established.
(1 = strongly disagree, 5 = strongly agree)
Transaction specific investments ($r = 0.67, p < .01$)
1. Our company had machinery and tools custom-made for working with our partner.
2. Our company can also use these machines for projects with other partners.
(1 = strongly disagree, 5 = strongly agree)

Note: The original questions are presented in the table. Prior to the empirical analysis, the scales for particular items were reversed to maintain construct consistency.

The research model includes two dependent variables: interorganizational trust and contract complexity. We used five items to measure interorganizational trust. Our definition characterizes interorganizational trust as a multicomponent construct based on three related components: forbearance from opportunism (measured by two items), care and concern (measured by two items), and lack of monitoring (measured by one item). The choice and specification of these items was based on earlier studies (de Jong & Klein Woolthuis 2008a, 2008b). Following Reuer and Ariño (2007), we measured contractual complexity by 13 contract clauses that indicated the level of detail in the legal framework of the alliance. The complexity of the contract was indicated by the number of clauses in the contract. The research model included two main variables: organizational cultural differences and interdependence. The former was measured by two items that assessed whether the respondent experienced any conflict resulting from potential differences in the organizational culture of the partnering companies. The latter was also measured by two items: one item assessed the dependence of the respondent firm and the second item the dependence of the partner. We included four control variables in our models. First, we include focal firm size as a variable to control for extraneous factors such as bargaining power and resource base (Yiang, Gao, & Li, 2008). Firm size was measured by two items: the number of employees and the turnover of the focal firm. Second, we included previous alliance history. This may influence the alliance governance, since past relationships can serve as basis for mutual understanding and, for instance, trust. Third, we included transaction-specific investments, because nonrecoverable investments increase relational risk and therefore the need to specify detailed contracts. Fourth, we included the reputation of the partner firm in our research model as reputation may substitute for governance mechanisms that act to verify the intentions and monitor the actions of business partners (Arend, 2009). Partner reputation was measured by two items. All of our multi-item constructs met the regular criteria for construct validity in terms of factor loadings (factor loadings large than .60), Cronbach's alpha (larger than 0.60 for three or more item constructs), or interitem correlation coefficients (positive and significant for two-item constructs).

Estimation Methods

Our research model consists of two models. In Model I we regress interorganizational trust on the main effects and the control variables. In Model II we do the same for contract complexity. We expected to find opposite effects for the main effects on interorganizational trust and contract complexity in Models I and II that, taken together, would offer evidence for the substituting effects of trust and contracts. We used an ordinary least square regression to estimate Model I. The measure for contract complexity ranges from 0 to 13, and therefore we used ordered logistic regression to estimate Model II (Reuer & Ariño, 2007). We estimated Models I and II separately, leaving refinements in estimation techniques, model specifications, and alternative measurements for robustness analyses. Occasionally, the data included missing values that were deleted prior to entering the variables into the regression.

EMPIRICAL RESULTS

The means, standard deviations, and correlations among composite indicators are shown in Table 1.2. The hierarchical OLS and ordered logistic regression results are provided in Table 1.3. The baseline specifications (Models 1a and 1b) present the control variables. Models 2a and 2b report the effects of the main variables on interorganizational trust and contract complexity, respectively. In preparation for the regression analysis, we performed the regular tests to obtain reliable estimates. The latter tests gave satisfactory results: neither heteroscedasticity nor non-normality was an issue. We tested for possible bias caused by collinearity among variables by calculating the variance inflation factor (VIF) for each of the regression coefficients. The VIF values were well below the cut-off value of 10 recommended by Neter, Wasseman, and Kutner (1985).

The various fit parameters show that our models increasingly fitted the data better. We restricted the discussion of the findings to the model that included main effects and control variables. Model 1b shows that organizational cultural differences indeed reduced trust as hypothesized ($\beta = -0.453$; $p < 0.01$). Hypothesis 1a was accepted. Model 1b also showed that interdependence increased trust but the parameter estimate was not significant ($\beta = 0.072$; n.s.). Hypothesis 1b was therefore rejected. Model 2b reports that organizational cultural differences lead to a more detailed contract. This confirmed our expectations but the estimate was not significant ($\beta = 0.036$, n.s.). Hypothesis 2a was therefore rejected. Model 2b shows that interdependence has a positive and significant effect on contract complexity ($\beta = 0.483$; $p < 0.01$). This is the opposite of what we hypothesized and to some extent aligns with Gulati and Singh (1998), who conclude that greater interdependence between alliance partners will induce collaborating firms to adapt to more hierarchical governance structures. The result, however, might also be related to the measurement of contractual complexity. We therefore performed an additional analysis.

A principal component factor analysis of the 13 contract clauses resulted in three different factors that are associated with three different contract functions. The first factor is interpreted as the traditional safeguarding function of a contract. Clauses loading high on this factor aim at safeguarding ownership and property rights, and limit the incentives towards opportunistic behavior by implementing clauses of technology ownership, methods, license agreements, patent rights, and confidentiality. The second factor corresponds to the coordination and commitment function of contracts, relating to clauses that concern

Table 1.2. Correlations, Means, and Standard Deviations (n = 391)[a]

	Mean	*SD*	*1*	*2*	*3*	*4*	*5*	*6*	*7*	*8*
1. Interorganizational trust	22.366	3.673	1.000							
2. Contract complexity	8.808	3.224	–0.001	1.000						
3. Interdependence	7.594	2.019	0.099	0.256	1.000					
4. Cultural differences	3.092	1.989	–0.521	0.039	0.037	1.000				
5. Firm size	6.711	2.293	0.051	0.050	–0.119	0.042	1.000			
6. Alliance history	7.194	3.230	0.161	–0.103	0.032	–0.113	–0.026	1.000		
7. Partner reputation	7.092	3.893	0.120	0.120	0.088	–0.165	0.005	0.178	1.000	
8. Transaction-specific investments	4.703	3.044	–0.037	–0.010	0.171	0.042	0.030	0.023	0.038	1.000

[a] Correlations larger than .10 are significant at 5%, and larger than .15 significant at 1%

Table 1.3. The Roles of Trust and Contracts in High-Tech Alliances

	Trust (Model 1a)	*Contracts (Model 2a)*	*Trust (Model 1b)*	*Contracts (Model 2b)*
Main effects				
Cultural differences			–0.453***	0.036
			(0.078)	(0.126)
Interdependence			0.072	0.483***
			(0.049)	(0.142)
Control variables:				
Firm size	0.067	0.157	0.036	0.157
	(0.062)	(0.134)	(0.060)	(0.140)
Previous history	0.102	–0.268*	0.054	–0.289*
	(0.064)	(0.144)	(0.058)	(0.149)
Reputation	0.152***	0.320**	0.088*	0.321**
	(0.057)	(0.129)	(0.049)	(0.135)
Transaction-specific investments	–0.008	–0.023	0.009	–0.129
	(0.053)	(0.132)	(0.050)	(0.139)
Constant	0.013		0.029	
	(0.057)		(0.052)	

*** $p < 0.01$, ** $p < 0.05$ and * $p < 0.10$

relationship management—the goal and outcome of the relationship—and hence reflecting a contract which focuses on the process of cooperation in the alliance. The third factor reflects a framework for unforeseen contingencies in the collaborative process. Contracts drawing primarily on these clauses function as safeguards against external contingencies by focusing on relationship adjustment, termination, and risk allocation. The results of the regression analysis for each of the three contract functions are presented in Table 1.4.

Table 1.4 largely confirms our Hypothesis 2b. Partners who are interdependent do not use a contract to safeguard risk. The effect of interdependence on this contract function is negative and significant ($\beta = –0.170, p < 0.05$). Interdependent partners also have less inclination to use contracts to safeguard from external contingencies. Here also the effect of interdependence on this contract function is negative and significant ($\beta = -0.299, p < 0.01$). Table 1.4 also reports a negative relationship between

Table 1.4. Contract Functions, Interdependence, and Cultural Differences in High-Tech Alliances

	Function 1 Safeguarding Risks and Spillover Contract	*Function 2 Coordination and Commitment Contract*	*Function 3 Safeguarding External Contingencies Contract*
Main effects			
Cultural differences	–0.024	–0.024	–0.014
	(0.057)	(0.062)	(0.0660
Interdependence	–0.170**	–0.072	–0.299***
	(0.071)	(0.060)	(0.062)
Control variables:			
Firm size	–0.066	–0.132**	–0.057
	(0.068)	(0.057)	(0.062)
Previous history	0.080	0.074	0.104
	(0.077)	(0.056)	(0.066)
Reputation	–0.068	–0.148**	–0.163**
	(0.074)	(0.069)	(0.063)
Transaction-specific investments	0.008	0.103	–0.002
	(0.074)	(0.064)	(0.062)
Constant	0.022	–0.008	–0.006

*** $p < 0.01$, ** $p < 0.05$ and * $p < 0.10$

interdependence and the coordination and commitment function of a contract, albeit this effect is not significant (β = -0.072, n.s.). Taking all results of Table 1.4 into account, we find convincing evidence that interdependence reduces a need to write complex contracts, provided that particular contract functions are accounted for. Table 1.4 confirms the nonsignificant relationship between organizational cultural differences and each of the contract functions reported for contractual complexity in Table 1.3. We therefore conclude that organizational cultural differences are not related to formal contracting mechanisms in Dutch high-tech alliances.

An interpretation of the results in Tables 1.3 and 1.4 offers support for our third hypothesis. We find that organizational cultural differences have opposite effects on the two governance mechanisms. Hence, the conclusion is that organizational cultural differences do have a different impact on trust and contracts as hypothesized—and, in so doing, support a substituting perspective. A similar conclusion holds for interdependence. Here we also find mutually excluding effects: Interdependence decreases the need for particular contract functions dedicated to safeguarding and it has a positive albeit not significant relationship with interorganizational trust. Finally, it is worthwhile to mention that our results hold while controlling for a large number of alternative determinants of trust or contract complexity in high-tech collaboration.

Robustness Analysis

We conducted further analyses to assess the robustness of our results. First, we substituted for missing values by using the mean value of a particular variable. This did not affect the results. Second, we used an alternative measurement for interdependence, measuring the distance between focal and partner dependence, which in our case had values of between –4 and +4. The value of 0 indicates that there is no difference between the alliance partners. This did not affect the results. Third, we included interaction terms between our main and dependent variables in order to more directly test the relationship between trust and contracts. For this, we extended the trust model by using contract complexity, as well as the interaction effects between contract complexity and cultural differences, and those between contract complexity and interdependence. In a similar vein, we extended the contract model using trust and the interaction effects between trust and the main effects variables. The results for these extended models show that none of the interaction terms were significant, thus adding to the confidence in our initial model specifications. Fourth, we used alternative estimation techniques that allowed us to estimate the trust and the contract models simultaneously. Seemingly unrelated regressions estimate equations simultaneously and account for correlation in the error terms of both model specifications. Structural equation modeling also estimates both models simultaneously. The results for the main effects are the same as reported in Table 1.3, by and large. Hence, in these supplementary analyses, the results were largely consistent with the initial results, adding to confidence in our findings vis-à-vis the role of contracts and trust in high-tech alliances.

CONCLUSIONS

The purpose of this chapter has been twofold. The first aim was to verify the importance of interdependence and organizational cultural differences in explaining the governance of strategic alliances. A review of the alliance literature suggests that the role of the alliance characteristics in the management of interfirm collaboration has received relatively little attention. They are acknowledged in alliance performance studies but seem to be largely ignored as explanatory variables for formal and informal mechanisms. Our theoretical considerations offer a first step in the explanation of whether and, if so, how interdependence and interorganizational cultural differences matter for the use of trust and contracts in high-tech alliances. We used our survey-based dataset of Dutch high-tech alliances to test our proposition. The empirical results offered convincing support for the hypothesized relationships. Among other things, we show that organizational cultural differences are detrimental for trust and hence hamper interfirm collaboration, given that trust generally is perceived to be beneficial for overall alliance performance.

The empirical findings also offer ample opportunities for the second aim of our research, that is, to offer an alternative perspective concerning the potential trade-off between formal and informal governance mechanisms. Prior alliance management studies report inconclusive findings with respect to the substitution of trust and contracts. We find mutually exclusive relationships with cultural differences as reducing trust (while not significantly impacting on contract complexity) and, simultaneously, interdependence as reducing the need for particular contract functions (while not significantly impacting on trust). It is worthwhile mentioning that these key findings hold for various tests of robustness. In our research setting we find evidence for the substitution between formal and informal governance forms.

Our research is not without limitations, which thus offers opportunities for future research. A first limitation concerns the static nature of our model. Some of the concepts used in the causal specifications are dynamic. In particular, the levels of trust and interdependence may change over time (whereas other elements such as organizational cultural differences and contract complexity are likely more stable alliance phenomena). This dynamic is not explicitly accounted for in our models. Future research may develop more dynamic perspectives on these alliance characteristics (by also accounting for the stage in the overall alliance lifecycle) and may determine how this can alter alliance governance mechanisms over time. Other research methods such as simulation techniques may offer complementary dynamic perspectives to the more static analyses presented here. A second limitation concerns the use of cross-sectional single-country data. It goes without saying that comprehensive datasets incorporating many different alliance characteristics are rare, the more so when alliance contract information is included. Companies are oftentimes reluctant to share detailed information on partner agreements. We show that such information can be collected in large-scale settings, thus enabling the analysis of simultaneous formal and informal governance mechanisms. That said, new information from other industries and other countries would make it possible to test the general validity of our cross-sectional and Dutch-based high-tech results. New data collection might well be especially directed towards Asian countries such as China, for which strategic alliances such as joint ventures are often a pre-requisite in order to successfully enter the Chinese market. It may also include documentation from companies that ally with firms in transition economies or those Central and Eastern European states where international collaboration is a result of government policies directed towards an increase in foreign direct investment and privatization in general.

REFERENCES

Arend, R. J. (2009). Reputation for cooperation: Contingent benefits in alliance activity. *Strategic Management Journal, 30*, 371-385.

Ariño, A., De la Torre, J., & Ring, P. S. (2001). Managing trust in corporate alliances. *California Management Review, 44*(1), 109-131.

Ariño, A., & Reuer, J. J. (2004). Designing and renegotiating strategic alliance contracts. *Academy of Management Executive, 18*(3), 37-48.

Artz, K. W., & Brush, T. H. (2000). Asset specificity, uncertainty and relational norms: an examination of coordination costs in collaborative strategic alliances. *Journal of Economic Behavior & Organization, 41*, 337-362.

Aulakh, P. S., Kotabe, M., & Sahay, A. (1996). Trust and performance in cross-border marketing partnerships: A behavioral approach. *Journal of International Business Studies, 27*, 1005-1032.

Baker, G., Gibbons, R., & Murphy, K. J. (2002). Relational contracts and the theory of the firm. *Quarterly Journal of Economics, 1*, 39-84.

Barney, J. B., & Hansen, M. H. (1994). Trustworthiness as a source of competitive advantage. *Strategic Management Journal, 15*, 175-190.

Beugelsdijk, S., Koen, C., & Noorderhaven, N. (2009). A dyadic approach to the impact of differences in organizational culture on relationship performance. *Industrial Marketing Management, 38*, 312-323.

Brockner, J., Siegel, P. A., Daly, J. P., Tyler, T., & Martin, C. (1997). When trust matters: The moderating effect of outcome favorability. *Administrative Science Quarterly, 42*, 558-583.

Chen, Y. (2000). Promises, trust, and contracts. *Journal of Law, Economics and Organization, 16*(1), 209-232.

Contractor, F. J. (2005). Alliance structure and process: Will the two research streams ever meet in alliance research? *European Management Review, 2*, 123-129.

Das, T. K., & Kumar, R. (2007). Learning dynamics in the alliance development process. *Management Decision, 45*, 684-707.

Das, T. K., & Kumar, R. (2010). Interpretive schemes in cross-national alliances: Managing conflicts and discrepancies. *Cross Cultural Management: An International Journal, 17*, 154-169.

Das, T. K., & Kumar, R. (2011a). Interpartner negotiations in alliances: A strategic framework. *Management Decision, 49*, 1235-1256.

Das, T. K., & Kumar, R. (2011b). Regulatory focus and opportunism in the alliance development processs. *Journal of Management, 37*, 682-708.

Das, T. K., & Teng, B. (1996). Risk types and inter-firm alliance structures. *Journal of Management Studies, 33*, 827-843.

Das, T. K., & Teng, B. (1998). Between trust and control: Developing confidence in partner cooperation in alliances. *Academy of Management Review, 23*, 491-512.

Das, T. K., & Teng, B. (2001). Trust, control, and risk in strategic alliances: An integrated framework. *Organization Studies, 22*, 251-282.

de Jong, G., & Nooteboom, B. (2000). *The causal structure of long-term supply relationships: An empirical test of a generalized transaction cost theory*. Dordrecht, Netherlands: Kluwer Academic Publishers.

de Jong, G., & Klein Woolthuis, R. J. A. (2008a). Contract research and high-tech alliances: Vistas for future studies. *Scandinavian Journal of Management, 24*, 284-289.

de Jong, G., & Klein Woolthuis, R. J. A. (2008b). The institutional arrangements of innovation: Antecedents and performance effects trust in high-tech alliances. *Industry and Innovation, 15*(1), 45-67.

de Jong, G., & Klein Woolthuis, R. J. A. (2009). The content and role of formal contracts in high-tech alliances. *Innovation: Management, Policy & Practice, 11*(1), 44-59.

de Jong, G., & Klein Woolthuis, R. J. A. (2010). The ex-post use of formal contracts in high-tech alliances. A contingency perspective. *Problems and Perspectives in Management, 8*, 48-60.

Denison, D. R., & Mishra, A. K. (1995). Toward a theory of organizational culture and effectiveness. *Organization Science, 6*, 204-233.

Dyer, J. H., & Chu, W. (2003). The role of trustworthiness in reducing transaction costs and improving performance: Empirical evidence from the United States, Japan, and Korea. *Organization Science, 14*, 57-68.

Guidice, R. M., & Cullen, J. B. (2007). A strong inference test of the effects of strategic interdependence on the JV control-performance relationship. *Journal of Business and Management, 13*(2), 127-153.

Gulati, R. (1995). Does familiarity breed trust? The implications of repeated ties for contractual choice in alliances. *Academy of Management Journal, 38*, 85-112.

Gulati, R., & Singh, H. (1998). The architecture of cooperation. Managing coordination costs and appropriation concerns in strategic alliances. *Administrative Science Quarterly, 43*, 781-814.

Gulati, R., & Sytch, M. (2008). Does familiarity breed trust? Revisiting the antecedents of trust. *Managerial and Decision Economics, 29*, 165-190.

Hill, C. W. L. (1990). Cooperation, opportunism, and the invisible hand: Implications for transaction cost theory. *Academy of Management Journal, 15*, 500-513.

Huber, G. P., & Power, D. J. (1985). Retrospective reports of strategic-level managers—Guidelines for increasing their accuracy. *Strategic Management Journal, 6*, 171-180.

Klein Woolthuis, R. J. A., Hillebrand, B., & Nooteboom, B. (2005). Trust, contract and relationship development. *Organization Studies, 26*, 813-840.

Krishnan, R., & Noorderhaven, X. M. N. G. (2006). When does trust matter to alliance performance? *Academy of Management Journal, 49*, 894-917.

Kumar, N., Scheer, L. K., & Steenkamp, J. E. M. (1995). The effects of perceived interdependence on dealer attitudes. *Journal of Marketing Research, 32*, 348-356.

Lei, D., Slocum, J. W. Jr, & Pitts, R. A. (1997). Building cooperative advantage: Managing strategic alliances to promote organizational learning. *Journal of World Business, 32*, 203-223.

Lowensberg, D. A. (2010). A "new" view on "traditional" strategic alliances' formation paradigms. *Management Decision, 48*, 1090-1102.

Lyons, B., & Mehta, J. (1997). Contracts, opportunism and trust: Self-interest and social orientation. *Cambridge Journal of Economics, 21*, 239-257.

Marcoulides, G. A., & Heck, R. H. (1993). Organizational culture and performance: Proposing and testing a model. *Organization Science, 4*, 209-225.

Mayer, R. C., Davis, J. H., & Schoorman, F. D. (1995). An integrative model of organizational trust. *Academy of Management Review, 20*, 709-734.

Mellewigt, T., Madhok, A., & Weibel, A. (2007). Trust and formal contracts in interorganizational relationships: Substitutes and complements. *Managerial and Decision Economics, 28*, 833-847.

Neter, J., Wasseman, W., & Kutner, M. H. (1985). *Applied linear regression models.* Chicago, IL: Irwin.

Nooteboom, B. (1996). Trust, opportunism and governance: A process and control model. *Organization Studies, 17*, 985-1010.

Nooteboom, B. (2002). *Trust: Forms, foundations, functions, failures and figures.* Cheltenham, UK: Edward Elgar.

Nooteboom, B. (2004). *Inter-firm collaboration, learning and networks: An integrated approach.* London, UK: Routledge.

Parkhe, A. (1998a). Understanding trust in international alliances. *Journal of World Business, 33*, 219-240.

Parkhe, A. (1998b). Building trust in international alliances. *Journal of World Business, 33*, 417-437.

Poppo, L., & Zenger, T. (2002). Do formal contracts and relational governance function as substitutes or complements? *Strategic Management Journal, 23*, 707-725.

Podsakoff, P. M., & Dennis, W. O. (1986). Self-reports in organizational research: Problems and prospects. *Journal of Management, 12*, 531-544.

Pothukuchi, V., Damanpoor, F., Choi, J., Chen, C. C., & Park, S. H. (2002). National and organizational culture differences and international joint venture performance. *Journal of International Business Studies, 33*, 243-265.

Razzaque, M. A., & Boon, T. G. (2008). Effects of dependence and trust on channel satisfaction, commitment and cooperation. *Journal of Business-to-Business Marketing, 10*(4), 23-48.

Reuer, J. J., & Ariño, A. (2007). Strategic alliance contracts: Dimensions and determinants of contractual complexity. *Strategic Management Journal, 28*, 313-330.

Sambasivan, M., Loke, S. P., Mohamed, Z. A., & Leong, L. C. (2011). Impact of interdependence between supply chain partners on strategic alliance outcomes: Role of relational capital as a mediating construct. *Management Decision, 49*, 548-569.

Smith, J. B., & Barkley, D. W. (1999). Selling partner relationships: The role of interdependence and relative influence. *Journal of Personal Selling and Sales Management, 19*(4), 21-40.

Veiga, J., Lubatkin, M., Calori, R., & Very, P. (2000). Research note measuring organizational culture clashes: A two-nation post-hoc analysis of a cultural compatibility index. *Human Relations, 53*, 539-557.

Williamson, O. E. (1985). *The economic institutions of capitalism.* New York, NY: Free Press.

Yang, H., & Lin, Z. (2005). Interfirm distances as channels for external resources: The choice between alliances and mergers & acquisitions. *Academy of Management Best Paper Proceedings*, BPS: I1-6.

Yiang, X., Gao, S., & Li, Y. (2008). How trust, dependence and prior ties influence the stability of alliances. *Journal of General Management, 34*(1), 1-17.

CHAPTER 2

KNOWLEDGE NETWORKS AND KNOWLEDGE FLOWS IN REGIONAL INNOVATION CLUSTERS

An Empirical Study of a High Technology Cluster In China

Jie Xue and Paul Olk

ABSTRACT

Previous research into regional innovation clusters has argued that high technology clusters exchange primarily technical knowledge in developing innovations. In this chapter, we propose that high technology innovation clusters will include technical knowledge but also may include two other types of knowledge. Drawing from market pull research, we examine the role of market knowledge and then we introduce the concept of managerial knowledge to regional innovation clusters research. We propose that the communication of different kinds of knowledge between organizations—through formal and informal ties—creates different knowledge flows that produce different

---

Strategic Alliances for Innovation and R&D, pp. 29–61

types of networks. To explore these arguments, we analyze whole network data for 56 electronic information firms in a Dongguang, China cluster. Specifically, we identify and illustrate these different knowledge networks for a focal company, analyze the resulting network structures for each type of knowledge, and explore which actors in the network contribute which type of knowledge. The research contributes to the knowledge-based view by elucidating the importance of different knowledge content to the enterprises' innovation and illustrates different knowledge flows in knowledge networks in the clusters.

INTRODUCTION

Research into industrial districts (Becattini, 1990; Brusco, 1990), clusters (Porter, 1990), innovative milieus (Camagni, 1991), regional innovation systems (Cooke, 2001) and learning regions (Asheim, 1996) have each addressed the importance of geographic regions as sources of innovation. In particular, researchers identify knowledge and knowledge spillovers as central to developing a regional competitive advantage and for general regional development (Asheim, Boschma, & Cooke, 2011; Asheim & Coenen, 2005). Knowledge spillover research has highlighted its localized nature (e.g., Alcácer & Chung, 2007; Almeida & Kogut, 1999; Bell & Zaheer, 2007; Gambardella & Giarratana, 2010; Jaffe, Trajtenberg, & Henderson, 1993; Owen-Smith & Powell, 2004; Sorenson, 2003), and noted that while knowledge flows across organizational boundaries, the intensity and effects of these flows are enhanced by spatial proximity (Jaffe, 1989; Gertler, 1995; Maskell & Malmberg, 1999). The focus on knowledge flow is important because, as Teece, Pisano, and Shuen (1997) argued, the knowledge breadth necessary to innovate and compete successfully can reside outside a firm's locus of expertise since its internal technological alternatives are "often rather narrow." Enterprises need to broaden their knowledge range through interactions with other enterprises. From a knowledge-based perspective, then, partnerships may be driven by complementarities (Das & Teng, 2000; Gulati, 1998). Firms can create synergistic combinations of assets, knowledge, and/or capabilities that may contribute to lowering the costs and/or increasing the perceived value of their offerings (Belderbos, Gilsing, & Lokshin, 2012).

The focus on knowledge and networks as sources of innovation is particularly important in technology industries (Ahuja, 2000; Owen-Smith & Powell, 2004; Powell, Koput, & Smith-Doerr, 1996; Stuart, Hoang, & Hybels, 1999). In these industries, interfirm collaboration has become common practice (Hagedoorn, 2002). Benefiting from the geographical proximity (Asheim, 1996; Cooke, 2001) and innovative culture (Saxenian, 1996) in high technology regions, it is more convenient to form interorganizational relationships within the region than in other places (Cooke, 2001). The formal cooperation (such as alliance, supplier-customer relationship) and the informal interactions based upon social relationships (such as face to face talking) can lead to the formation of regional organizational networks. Therefore, interorganizational relationships can serve as heterogeneously distributed repositories of knowledge, and interorganizational interactions increasingly produce and diffuse knowledge in the cluster. A knowledge network, then, is formed from the knowledge transfer among organizations in the cluster connected by different social relationships.

Although it has been demonstrated that interaction among different organizations and the formation of a knowledge base can be (highly) beneficial for firms in clusters (Cooke, 2001; Porter, 1990; Saxenian, 1996), the knowledge content in the base, the formation of knowledge networks and different sources of knowledge flows in clusters have not been researched. Studies of the knowledge bases of the cluster have largely focused on technological knowledge (e.g., Asheim & Coenen, 2005; Asheim, Coenen, Moodeysson, & Vant, 2007; Asheim & Gertler, 2005) or on the knowledge properties, such as tacit knowledge or codified knowledge (e.g., Nonaka & Takeuchi, 1995). For example, researchers studying technological knowledge complementarities (e.g., Ahuja, 2000; Ernst, Lichtenthaler, & Vogt, 2011; Grimpe & Kaiser, 2010; Owen-Smith & Powell, 2004; Teece, 1992) typically ignore other knowledge content such as market knowledge and managerial knowledge, even though these are essential elements in the innovation process (Agarwal, Echambadi, Franco, & Sarkar, 2004; Birkinshaw, Hamel, & Mol, 2008; Marvel & Lumpkin, 2007). While some studies (e.g., Burgers, Van Den Bosch, & Volberda, 2008; Marvel & Lumpkin, 2007) recognized more than one type of knowledge in regional networks, research has generally not empirically examined the existence of various knowledge types and the interdependent relationships among them.

To address these issues, we study an electronic information cluster in China using social network analysis (SNA). SNA has emerged as a particularly appealing tool to regional scientists because it offers options for investigating structural and spatial patterns of knowledge-intensive networking (Knoke & Yang, 2008; TerWal & Boschma, 2009). We believe our study contributes to the research on regional innovation clusters in several ways. First, we help develop a clearer understanding of the types of knowledge that are shared among actors in a regional innovation cluster. We add to the traditional distinction between technological knowledge and market knowledge (e.g., Freeman, 1974; Henderson & Clark, 1990; Rosenberg, 1982; von Hippel, 1976) the concept of managerial knowledge, which is composed of management practice, process, structure, and technique that also contribute to long-term firm success (Birkinshaw et al., 2008). Second, the three different types of knowledge content helps us explain the formation and structure of knowledge networks in regional innovation clusters, and understand the interdependence and complementarities of different knowledge content in the process of innovation. Third, our research demonstrates the presence of different kinds of knowledge networks and knowledge flows among actors in a high-technology cluster. The analysis helps to understand diversity in knowledge networks and knowledge sources in innovation clusters and to comprehend the differences among various knowledge flows. Finally our study, based on a high-technology cluster in south China, adds empirical evidence to the literature on regional innovation clusters. Most research has focused on successful clusters in the United States, Europe, Japan or South Korea (e.g., Asheim & Coenen, 2005; Porter, 1990; Saxenian, 1996). Under the trend of globalization, firms in regions in developing countries such as China also try to develop competitive advantages through cooperation and learning. Extending data collection to a new context helps broaden the applicability of knowledge-based theory.

The chapter is structured as follows. First, we draw from regional innovation clusters, knowledge transfer in alliances and knowledge networks research to develop a set of hypotheses. After discussing our methodology, we present and discuss our empirical findings. Finally, we highlight the main conclusions of the study, and point out avenues for further theoretical and empirical work in this emerging area of research.

THEORY AND HYPOTHESES

Knowledge Bases and Knowledge Content in Clusters

'The literature on learning and knowledge transfer in clusters primarily addresses the tacit and/or implicit nature of knowledge (Asheim, 1996; Cooke, 2001; TerWal & Boschma, 2011). For example, in three differentiated regional knowledge bases brought forward by Asheim and colleagues (Asheim & Coenen, 2005; Asheim et al., 2007; Asheim & Gertler, 2005), they identified codified and tacit knowledge as two fundamental knowledge types. They then categorized the knowledge in an innovation region into science-based (analytical) knowledge, engineering-based (synthetic) knowledge and arts-based (symbolic) knowledge. They did not, however, address the diversity of knowledge content, and the likely outcome of various knowledge networks and the potentially unequal knowledge flows in clusters. Because knowledge is a multifaceted construct (Burgers et al., 2008; Shane, 2000), such models may be underspecified (Sullivan & Marvel, 2011).

Research on knowledge content suggests that both technology and market knowledge have important implications for firms (Agarwal et al., 2004; Marvel & Lumpkin, 2007). Innovation entails combining technical information or knowledge (which embodies the "solution" knowledge) with knowledge of user needs and translating the combination into new products that the users want (Bogers, Afuah, & Bastian, 2010). This means that unless firms have proper market knowledge they cannot create innovations that also meet market needs. Producers can take advantage of users as innovators, and research has explored how users can be active innovation participants as co-producers of products, services, or value in general (Buur & Matthews, 2008; Prahalad & Ramaswamy, 2003; Vargo & Lusch, 2004). Hence, technological knowledge and market knowledge are indispensable drivers of the innovation process. Technological knowledge refers to the knowledge about products, technologies, and/or processes that an enterprise possesses that is relevant to its business (Burgers et al., 2008), and it facilitates skill development and competencies that help achieve competitive advantages (Dodgson, 1993). Market knowledge refers to the knowledge an enterprise possesses about potential customers, distribution channels, and how the market functions that is relevant to its business (Burgers et al., 2008). Market knowledge provides enterprises the ability to serve customers appropriately because it provides insights into customers' preferences, effective distribution channels, and effective manufacturing procedures (Danneels, 2002). It also helps firms explore and exploit innovative opportunities more efficiently (Li & Calantone, 1998).

Beyond these two types of knowledge, firms trying to innovate may be faced with the need to renew themselves. Beyond offering new products and services they may need to change the nature of management within the organizations (Vaccaro, Jansen, Van Den Bosch, & Volberda, 2012). Crossan and Apaydin (2010) argued that it takes time, investment, and managerial effort for a company to survive the continuous changes in the environmental and competitive landscape. Like technical and marketing knowledge, managerial knowledge is also fundamental to firm survival and growth. Managerial knowledge entails the way in which management work is performed, which refers to the knowledge about management practice, process, structure, and/or technique an enterprise possesses that is intended to further organizational goals (Birkinshaw et al., 2008; Teece, 2007). Numerous cases have demonstrated that appropriate managerial knowledge could help companies gain competitive advantages, such as GM's M-form organization structure, Motorola's Six Sigma quality control methodology, Toyota's Lean Production System and P&G's Brand Management (Mol & Birkinshaw, 2009). Under the direction of proper managerial knowledge, innovative firms can create an internal context to promote innovation activity by providing support for experimentation

(King, Anderson, & West, 1992; West & Anderson, 1992), by cultivating an innovation culture (Madjar, Oldham, & Pratt, 2002; King et al., 1992; West & Anderson, 1992), by supporting learning and development of employees, by fostering the acceptance of diversity within the group (Crossan & Hulland, 2002), by managing the R&D activities efficiently (Bard, Balachandra, & Kaufmann, 1988), and by boosting the cooperation with outside partners (Schilke & Goerzen, 2010). What is unique about managerial knowledge is that it typically emerges without a dedicated infrastructure (such as research labs—which aid technical innovation), and is relatively abstract and intangible, which makes it potentially complex and ambiguous (Birkinshaw et al., 2008). Table 2.1 provides a summary of the definition and examples of the three knowledge types.

Table 2.1. The Definition and Examples of Different Knowledge Types and Knowledge Flows

Knowledge Type	*Definition*	*Example of Knowledge Flow*
Technological knowledge	Knowledge about products, technologies, and/or processes that an enterprise possesses that is relevant to their business.	If firm A provides specific technology solution to firm B in their cooperation, then we can tell there is technological knowledge flow from A to B, and firm B should determine the value of technological knowledge they get from firm A.
Managerial knowledge	Knowledge about management practice, process, structure, and/or technique an enterprise possesses that is intended to further organizational goals.	If firm A get knowledge about project management and quality management from firm B through their cooperation, then we can tell there is managerial knowledge flow from B to A,and firm A should determine the value of managerial knowledge they get from firm B.
Market knowledge	Knowledge about potential customers, distribution channels, and how the market functions that is relevant to their business.	Firm A provides CPU to firm B to produce personal computer in their cooperation. Firm B acknowledges that customers like CPU with higher energy saving capability, and feeds back customers' demand to firm A, then we can tell there is market knowledge flow from B to A, and firm A should determine the value of market knowledge they get from firm B.

Research suggests that innovation development requires persistent knowledge accumulation pertaining not only to technology, but also to the market (Danneels, 2002; Leonard-Barton, 1988). We also believe that more managerial knowledge could contribute to long-term firm success. Therefore, we propose that the knowledge base in regional clusters is composed of technological knowledge, market knowledge and managerial knowledge.

Hypothesis 1: *There are three different types of knowledge content in knowledge bases in high technology industrial clusters.*

Knowledge Networks and Knowledge Flows in Clusters

Networks are considered to be a highly efficient arrangement for sourcing innovation-related knowledge (Powell, 1990). The potential for interorganizational learning, which is knowledge transfer across organizational boundaries (Dyer & Nobeoka, 2000), is a key advantage of strategic alliances (Goerzen & Beamish, 2005). The capability to transfer knowledge effectively from the alliance partner also plays a central role for success (Mowery, Oxley, & Silverman, 2002; Teece, 2007). In regional innovation clusters such ties are facilitated by geographic proximity. The resulting organization's networks play a particularly important role in geographically concentrated industries because the likelihood and effects of ties are often constrained by distance (Whittington, Owen-Smith, & Powell, 2009).

A large and growing body of empirical research shows that social relationships and the networks these relationships constitute are influential in explaining the processes of knowledge creation, diffusion, absorption, and use. The acquisition and exploitation of knowledge are often social processes (Kogut & Zander, 1992; Yli-Renko, Autio, & Sapienza, 2001). Firms in the region may acquire knowledge through interactions with the external environment via their network partners (Kaish & Gilad, 1991). Phelps, Heidl, and Wadhwa (2012) conceptualized knowledge networks as consisting of *nodes* that serve as repositories of knowledge and agents that search for, adopt, transmit, and create knowledge. Nodes are simultaneously sources and recipients of information and knowledge. We define the concept of a knowledge network as a set of actors connected by knowledge communications, which are the foundation of an innovation network and the knowledge base of enterprise innovation. A knowledge network can help a company acquire external knowledge and competencies in order to strengthen and renew its competitive advantage, contributing to the firms' dynamic capabilities (Armstrong & Shimizu, 2007; Eisenhardt & Martin, 2000).

Drawing from the resource-based view (RBV), numerous researchers have emphasized technology as the key driver of innovation (Hitt, Bierman, Shimizu, & Kochhar, 2001; Hoskisson, Hitt, Wan, & Yiu, 1999). Additionally, though knowledge networks research spans multiple fields and levels of analysis including knowledge sharing between individuals (Bouty, 2000), the influence of social network structure on the diffusion and the adoption of innovations (Bothner, 2003), how networks influence knowledge production and diffusion (Azoulay, Zivin, & Wang, 2010; Jackson, 2008), how strategic alliances affect interfirm knowledge transfer (Lane & Lubatkin, 1998) and how alliance network structure affects firm innovation (Ahuja, 2000; Schilling & Phelps, 2007) diversity of knowledge content in knowledge network has not received much research attention. As noted, more and more innovation takes inspiration and knowledge from the demand side (Lettl, Herstatt, & Gemuenden, 2006; Nambisan & Baron, 2010; Priem, Li, & Carr, 2012). Market knowledge can be a major driver of innovation, especially in consumer goods sectors, with some firms intentionally involving consumers in their innovation processes (e.g., Franke & Shah, 2003; von Hippel, 2005). In essence, relying on a network for market knowledge acquisition allows for rapid access to knowledge, more accurate knowledge regarding market preferences, and a higher likelihood of finding novel ways that market needs may be met (Sullivan & Marvel, 2011). Moreover, companies differ considerably in their routines and managerial approaches and some companies may learn management skills from others when they interact in the networks. That is, managerial knowledge learning may occur.

Therefore, we introduce different knowledge content into the research of knowledge networks, and believe there are three different knowledge networks and knowledge flows in regional innovation clusters. Following these arguments leads us to our formal hypothesis:

> **Hypothesis 2a:** *There are three different knowledge networks and knowledge flows in high technology industrial clusters.*

According to the aforementioned research, various knowledge content has different functions in the enterprise's innovation process, but most innovation network research has not associated differences in knowledge content with different types of partners. Instead, researchers typically focus on the acquisition and development of technology knowledge because it relates to an entrepreneur's ability to create products that meet market demands (Clarysse, Wright, & Van de Velde, 2011), to build competencies (Danneels, 2002) and to achieve desired outcomes (Zahra, Ireland, & Hitt,

2000), like developing innovations. Similarly, most explanations for the widespread use of collaborative arrangements by firms engaged in R&D intensive activities build from an interest in the dispersion and rapid development of the scientific knowledge, skills, and resources (Powell, White, Koput, & Owen-Smith, 2005). Further, Sullivan and Marvel (2011) argued that knowledge network reliance upon technology knowledge is likely to be particularly important for developing innovations because of the challenges associated with keeping up with these rapid changes in technology.

Several research streams suggest that a firm's access to varying knowledge content is important for its innovation. Alliance portfolio research has focused on potential complementarities between different partner types, as each brings different sets of knowledge or complementary capabilities (Lavie, 2007; Vassolo, Anand, & Folta, 2004). Burgers et al. (2008) suggested that technology firms often overlook the acquisition of market knowledge which can negatively affect the successful commercialization of new products because of the fast-paced, disruptive, and dynamic market environments within which they operate (Agarwal et al., 2004; Burgers et al., 2008). Priem et al. (2012) argued that successful innovations can be consumer driven rather than resource or technology driven, and consumer knowledge can play a key role in entrepreneurial idea discovery. Further, a firm's interaction with a supplier and/or customer and the interaction with a competitor may have different effects on firm's objective and performance, with the latter frequently focusing on more radical innovations and the former on cost reduction or on reducing time to market (Belderbos, Carree, & Lokshin, 2004; Tether, 2002). The research on the impact of managerial knowledge on innovation is not abundant, but we still can comprehend its importance from the case of GlaxoSmithKline (GSK) PLC, the U.K. pharmaceuticals giant. The company was suffering from declining productivity in its research and development activities in 2000. It found that the flexibility and responsiveness of biotech firms is essential to the success of a pharmaceutical innovation. The company then radically restructured its drug development operations into seven Centres of Excellence. These were designed to have biotech-like levels of flexibility and autonomy while still benefiting from the scale of GSK's global presence (Birkinshaw & Mol, 2006). The new practice named Centres of Excellence for Drug Discovery was a new managerial knowledge implemented in GSK, which helped its R&D activities. From these points, we conclude that varying knowledge content may have different impact on enterprise's innovation.

Further, the knowledge flow in different knowledge networks should be heterogeneous and result in different network structures. A firm's collaboration with different types of partners is driven by different motives and characterized by different risks and corresponding needs for control (Parkhe, 1993). This indicates the varying importance of different knowledge content in the interaction between partners. Depending on their deficiency in terms of technological, managerial and market knowledge, companies in clusters could consider and search for the specialized knowledge which is held by other organizations in the cluster, and simultaneously think about how to gain access to the knowledge. Our arguments can be summarized in a hypothesis:

Hypothesis 2b: *Different knowledge flows in different knowledge networks are unequal in amount, density and transitivity in high technology industrial clusters.*

Knowledge Flows and Enterprises Innovation in Clusters

From a knowledge perspective, partnerships are driven by resource complementarities across partner types (Das & Teng, 2000; Gulati, 1998). The extant empirical literature on knowledge sourcing suggests that firms rely heavily on knowledge exchange with external parties, such as suppliers, customers, universities, other key individuals, and sometimes even competitors (Chesbrough, 2003; Laursen & Salter, 2006; von Hippel, 2005). Sullivan and Marvel (2011) argued that by sourcing knowledge from customers and/or other external resource providers like suppliers, competitors, research institutes, universities and professional services, the enterprise is able to develop a more accurate understanding of how customers can effectively be reached. The focal company can access market knowledge from various partners. For example, some professional services, such as venture capital firms, along with specialized law firms, act as sources of managerial and financial expertise (Powell, Koput, Smith-Doerr, & Bowie, 2002; Patton & Kenney, 2005). Moreover, public research organizations (PROs) including universities, research institutes, and teaching hospitals are vital sources of innovation because they are the producers of both technical personnel and cutting-edge scientific knowledge (Whittington et al., 2009). In short, different partner types play different roles in complementing a firm's own resources and capabilities, related to potentially different goals of the alliances (Belderbos et al., 2004; Faems, Van Looy, & Debackere, 2005).

Although research has demonstrated the benefits of collaboration with customers and/or suppliers as well as with competitors (e.g., Ahuja, 2000; Frohlich & Westbrook, 2001), the source of specific knowledge content in knowledge networks has not been studied. Belderbos et al. (2012) distinguished between alliances with suppliers and/or customers (vertical alliance) and alliances with competitors (horizontal alliance), but the research focus is mainly on the context of technological cooperation. Likewise, in the literature on alliances, most studies have not explicitly discussed the knowledge content provided by different kinds of partners (Ahuja, 2000; Rowley, Behrens, & Krackhardt, 2000; Sampson, 2007). In contrast, the supply chain literature has largely focused on vertical alliances with suppliers or customers and restricted the knowledge type mainly to technological knowledge (Rosenzweig, Roth, & Dean, 2003; Vickery, Droge, & Calantone, 2003). The objectives and performance effects of vertical alliances with suppliers or customers have also been found to differ from those of horizontal alliance with competitors (Belderbos et al., 2004; Tether, 2002). We propose, therefore, that different actors in regional innovation clusters contribute different types of knowledge.

> **Hypothesis 3a:** *Different actors—that is, competitors, suppliers, customers, universities, industry associations and public technology platforms—in clusters contribute different types of knowledge flows in high technology industrial clusters.*

The Knowledge-Based View (KBV) considers the firm as "a knowledge-integrating institution" (Grant, 1996) or "a social community specializing in the speed and transfer of knowledge" (Kogut & Zander, 1996). Research suggests new knowledge is created from the novel combination of existing knowledge (Fleming, 2001). In addition, creativity research has suggested that accessing diverse pools of knowledge and developing skills to establish novel linkages among them are important conditions to generate creative outcomes (Simonton, 1999). It is the "package nature" of knowledge that makes capabilities difficult to buy, imitate, assemble, substitute, or replace (Hunt, 1999).

As noted, innovation research has found that combining technical information with the knowledge of user needs is helpful for creating new products the users want (Bogers et al., 2010). Similarly, the capability to effectively transfer technical knowledge from network partners and to implement project management effectively—a kind of managerial knowledge plays a central role the success of innovation (Mowery et al., 2002; Teece, 2007). Consequently, although research may not always recognize it, the three types of knowledge content, while distinct, may in practice be highly related. Therefore we predict that:

Hypothesis 3b: *According to the knowledge integration requirement of innovation, the three kinds of knowledge networks are positively correlated in high technology industrial clusters.*

METHOD AND ANALYSIS

Setting and Data Collection

The research setting for this study was an electronic information industry cluster in Dongguan, China. Dongguan is located in the Pearl River Delta in south China, which is the world biggest manufacturing base of electronic information products, and the most developed region in China. Firms in this industry cluster research produce and market hardware and software that relate to electronic products, computer products and mobile devices. According to statistics of the Dongguan government, there are more than 3,500 electronic information enterprises in the cluster, and the local self-supply rate of a computer is up to 90%. That is, the electronic information industry cluster is capable of providing all parts for a computer except the CPU. In technology industries like electronic information, scientific and related technological developments evolve rapidly (Orsenigo, Pammolli, & Riccaboni, 2001; Powell et al., 2005), generating uncertainty for firms about which technological opportunities to seize, and resulting in a dispersion of technological knowledge and competences required for conducting research and development across different firms (Powell et al., 1996). Hence, the electronic information industry cluster in Dongguan provides us a prime setting to study knowledge networks.

For our data, we started with a list of the 100 best enterprises from all of the enterprises in the region, based upon the firms' innovation and production capabilities, put together by the Dongguan government. With the help of the Guangdong Institute of S&T Information, we contacted to 91 in the 100 enterprises by phone. Eighteen enterprises indicated that they have no cooperation with local enterprises and 17 enterprises told us they are unavailable to take part in the survey, resulting in 56 enterprises that agreed to participate in our study. We also included in our questionnaire the name of six universities in the Guangdong province that emphasize cooperation with electronic information firms. Additionally, the area has industry associations and public innovation platforms devoted to servicing electronic information firms in the Pearl River Delta region. We added one industry association option and one public innovation platform option to the questionnaire. This resulted in a total of 56 enterprises, 6 universities, 1 industry association option and 1 public innovation platform option listed in the network questionnaire.

Guangdong Institute of S&T Information helped us send the questionnaire by e-mail to an informant in each firm. The informants were usually the director of the firm's executive office or the assistant to general manager, who knows the company's formal and/or informal cooperation or interactions with other firms, universities, industry associations and public innovation platforms very well. These individuals also have contact with key managers in the firm if help was needed finishing the questionnaire. We received all 56 questionnaires back. We found some had omitted answers or provided inconsistent answers, so we called some informants to confirm the results, leading to 56 valid questionnaires.

Measures and Analytical Approach

We designed the questionnaire to obtain data for our four variables collaborative relationship, technological knowledge flow, market knowledge flow, and managerial knowledge flow. For collaborative relationships, we examined formal and informal cooperation. Formal, contract-based interorganizational ties are largely ineffective in providing external knowledge relative to informal, interpersonal research collaborations that span organizational boundaries (Liebeskind, Oliver, Zucker, & Brewer, 1996). Likewise, Phelps, Heidl, and Wadhwa (2012) suggested that examining both formal and informal relationships at multiple levels of analysis is important for understanding better how knowledge networks influence organizational knowledge outcomes. We therefore defined collaborative relationship as a relational and binary variable. The company informant noted whether there was any informal or formal collaboration with each of the other 56 companies as well as the universities, industry association or public innovation platform. Thus, the variable collaborative relationship has a value of 1 if the two enterprises have had any formal or informal cooperation in the last three years and a value of 0 if they have not cooperated. The three knowledge flow types were evaluated for value and direction. The informant noted what kind of knowledge the firm received from the cooperation and identified the value of the specific knowledge they acquired, ranging from 0 (get nothing), 1 (a bit valuable knowledge) to 3 (very valuable knowledge). Lastly, the informant selected the type of firm: competitor, supplier or customer.

We used Social Network Analysis (SNA) to analyze the data. This analysis focuses on the abstract contours of a social space comprising concrete relationships among entities. Although SNA has largely ignored physical space, the idea that place is important to understanding innovation has wide credence (Whittington et al., 2009). Moreover, evidence supports the effect of proximity on network tie formation (Powell et al., 2005; Sorenson & Stuart, 2008). Integrating social and physical space holds promise for understanding knowledge networks and knowledge flows in a region.

Analysis and Results

The results from the social network analysis provide support for our hypotheses. Figures 2.1 to 2.4 show the cooperation network and the three kinds of knowledge networks, respectively. The notable disparities among the figures support our hypothesis 1, which claimed that there are three different types of knowledge content in high technology industrial clusters. From the figure, we can directly perceive that the cooperation network and the technological knowledge network are similar, and the technological knowledge network is different from the market knowledge network and managerial knowledge network. In the technological knowledge network, all of the knowledge transfers between firms in the cluster are bidirectional, which means companies get technological knowledge from each other in the technological cooperation pairs. The knowledge transfer is different in market and managerial knowledge networks, indicated by the unidirectional ties. Some firms get knowledge from the partner, but not vice versa. Hence, the graphs suggest the technological knowledge network is the foundation of cooperation between firms in the cluster. This may indicate that acquiring more technology knowledge is important for building other competencies (Danneels, 2002), and regional clusters depend on the networks that arise from reciprocal linkages among co-located organizations (Whittington et al., 2009). In addition, the degree centrality of Public Technology Platform (PTP, no. 63) and Industry Association (IA, no. 64) are bigger than other nodes in the cooperation network, illustrating that the PTP and IA in this cluster are very important information hubs for firms.

Hypothesis 2a and 2b, stating there are three different knowledge networks and knowledge flows, and different knowledge flows in different knowledge networks are unequal in high technology industrial clusters, are supported by several whole network indexes presented in Tables 2.2 and 2.3. The Amount of Knowledge Flow among the three different knowledge networks is different. The amount of technological knowledge (416) is much more than market (173) and managerial knowledge (201), highlighting the importance of acquiring more technology knowledge to build competencies (Danneels, 2002) and achieve desired outcomes (Zahra et al., 2000).

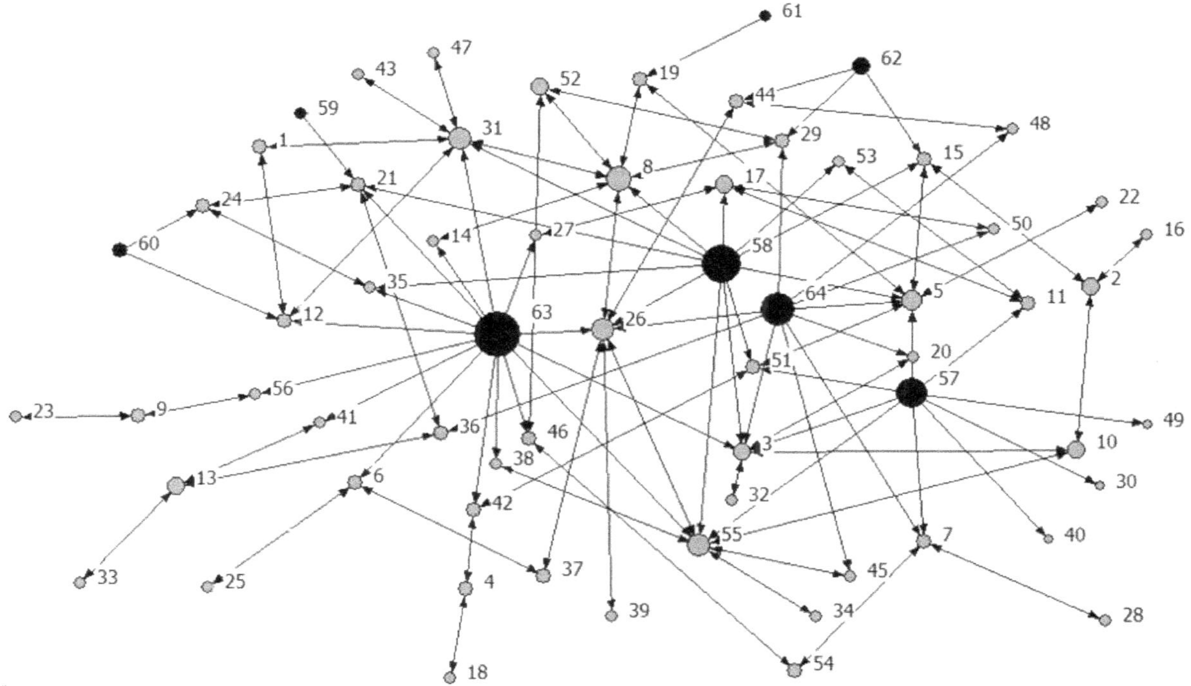

Figure 2.1. Cooperation network.
Grey circle: firms.
Black circle: universities (57-62), public technology platform (63), industry association (64).
The size of the circle depends on the degree centrality of the node.

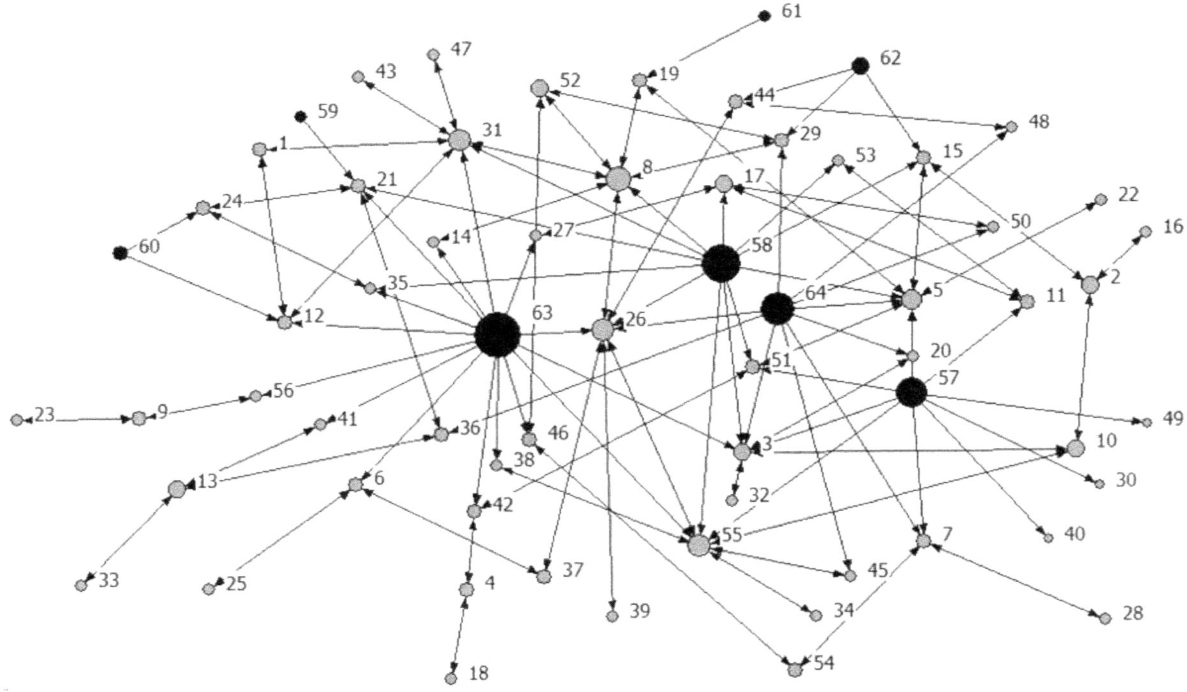

Figure 2.2. Technological knowledge network.
Grey circle: firms.
Black circle: universities (57-62), public technology platform (63), industry association (64).
The size of the circle depends on the out-degree centrality of the node, hereinafter inclusive.

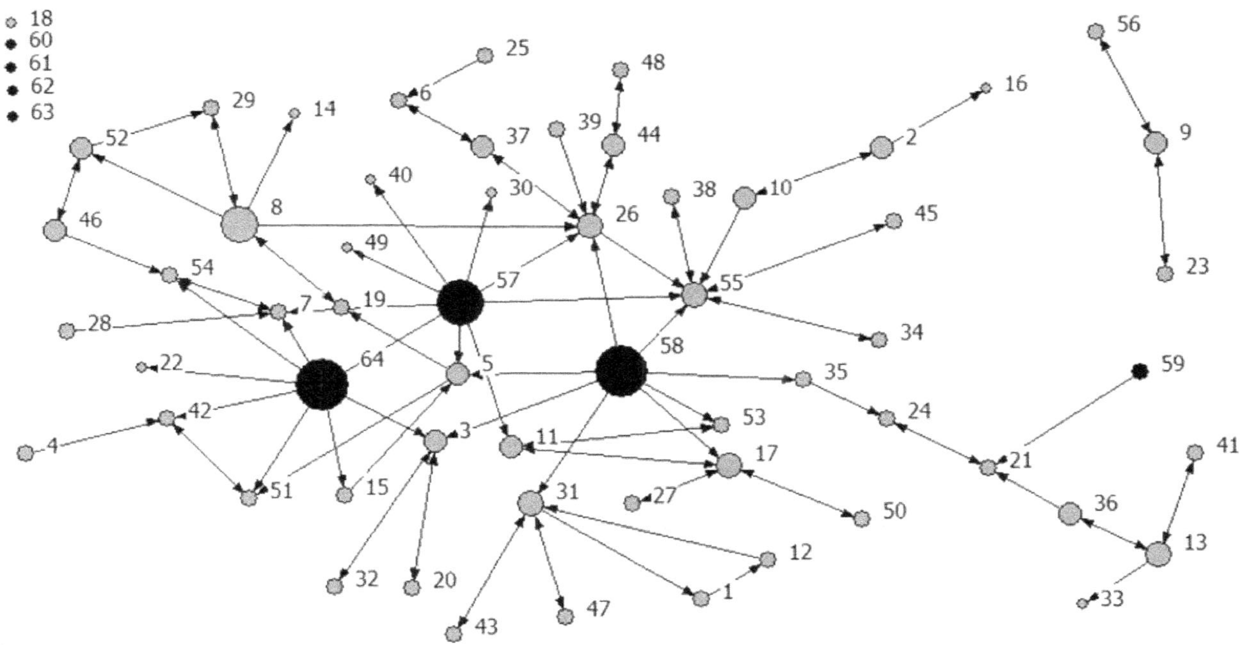

Figure 2.3. Managerial knowledge network.
Grey circle: firms.
Black circle: universities (57-62), public technology platform (63), industry association (64).
The size of the circle depends on the out-degree centrality of the node, hereinafter inclusive.
The five independent circles (18, 60, 61, 62, and 63) means organization has no managerial knowledge exchanges with other organizations in the cluster.

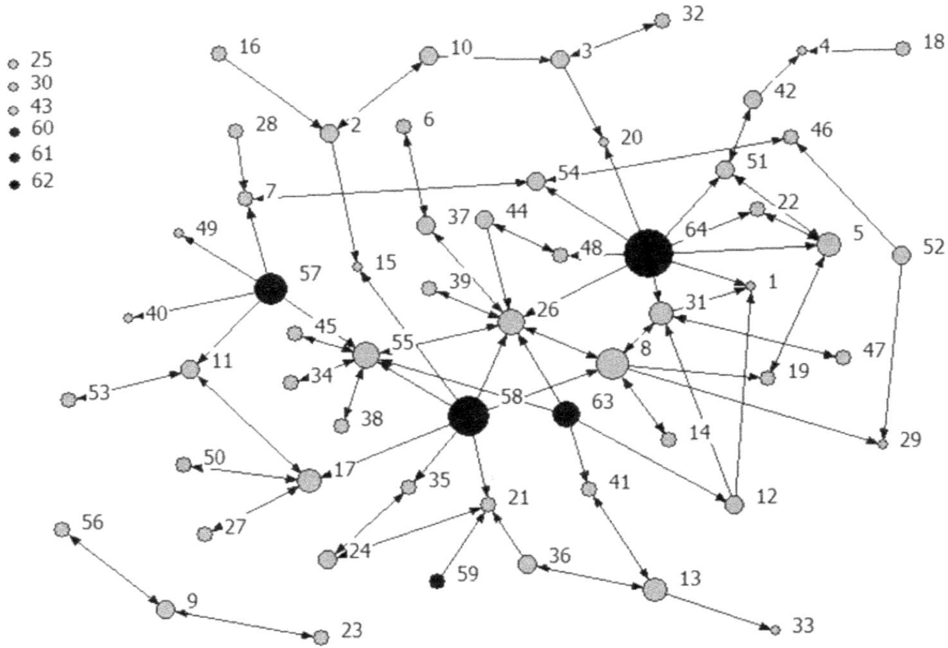

Figure 2.4. Market knowledge network.
Grey circle: firms.
Black circle: universities (57-62), public technology platform (63), industry association (64).
The size of the circle depends on the out-degree centrality of the node, hereinafter inclusive.
The six independent circles (25, 30, 43, 60, 61 and 62) means organization has no market knowledge exchanges with other organizations in the cluster.

Table 2.2. Whole Network Structural Indexes of Networks

	Cooperation Network	*Technological Knowledge Network*	*Market Knowledge Network*	*Managerial Knowledge Network*
Direction of ties	Undirected	Directed	Directed	Directed
Value of ties	–	0-3	0-3	0-3
Number of Observations	64	64	64	64
Density	0.041	0.103	0.043	0.050
Overall graph clustering coefficient	0.084	0.199	0.091	0.175
Degree Network Centralization (%)	22.12	17.750 (Outdegree) 7.373 (Indegree)	10.539 (Outdegree) 7.262 (Indegree)	11.393 (Outdegree) 8.116 (Indegree)
Betweenness Network Centralization (%)	21.17	21.29	5.40	2.62
Transitivity (%)	13.79	10.56	8.45	5.26
Number of weak components(with 3 nodes at least)	1	1	2	2
Proportion of nodes in the biggest component (%)	100	100	85.9	87.5
Fragmentation (prop. of nodes that cannot reach each other) (%)	0	0	26.2	23.5
The amount of knowledge flow†	–	416	173	201

† Measured by the sum of Degree Centrality.

The measures of Network Density and Clustering Coefficients capture the degree of connectivity in a network, and for our sample suggest strong differences among the networks. Extant literature suggests network density increases knowledge transfer among network contacts and enhances learning (Morgan & Soerensen, 1999; Morrison, 2002; Reagans & McEvily, 2003), and network connectivity (density) increases the rate, extent, and fidelity of information diffusion in a network (Singh, 2005).

Table 2.3. Correlations and Densities Comparison Between Different Knowledge Networks

	Densities Comparison †	*QAP Correlations ‡*
Technological knowledge network vs. Managerial knowledge network	4.6043*	0.670***
Technological knowledge network vs. Market knowledge network	5.1237*	0.660***
Managerial knowledge network vs. Market knowledge network	0.8369	0.584***

*$P < 0.05$; ***$P < 0.001$; †Bootstrap Paired Sample T-test; ‡QAP Statistics.

This can increase an innovation's diffusion by spreading the information about it (Abrahamson & Rosenkopf, 1997) and improving the network members' innovativeness (Ebadi & Utterback, 1984). These two indexes are different among the three knowledge networks, with technological knowledge network (0.103/0.199) having more dense ties than either market (0.043/0.091) or managerial (0.050/0.175) knowledge networks. At the same time, the difference of density between technological and market knowledge networks (*t*-test value 5.1237), and between technological and managerial knowledge networks (*t*-test value 4.6043) are significant at 95% confidence interval. The number of components and the transitivity indexes indicate that knowledge transitivity in the technological knowledge network is much better than in market and managerial knowledge networks. The transitivity index in technological knowledge network (10.56) is larger than in market (8.45) or managerial knowledge networks (5.26), and there is only one component in the technological knowledge network instead of two in market and managerial knowledge networks.

We also employed two Network Centralization indexes Degree and Betweenness Network Centralization to illustrate the extent of diversity of nodes in knowledge networks. These show that the technological knowledge network (17.750/21.29) is quite different from market (7.262/5.40) and managerial knowledge networks (8.116/2.62). In addition, we use Degree Centrality and Betweenness Centrality indexes to measure the power position and information control capability of a single node in different networks. Greater centrality is associated with timelier access to more, richer, and more diverse information and knowledge, increasing the extent to which actors learn from their network and their potential to synthesize and recombine this information and knowledge into novel ideas (e.g., Burt, 2004; Ebadi & Utterback, 1984; Morrison, 2002). Table 2.4 contains detailed information about these indexes. There are 10 (outdegree) and

Table 2.4. Degree Centrality and Betweenness Centrality of Different Knowledge Networks

	Degree Centrality†						*Betweenness Centrality‡*		
	Technological Knowledge Network		*Market Knowledge Network*		*Managerial Knowledge Network*		*Technological Knowledge Network*	*Market Knowledge Network*	*Managerial Knowledge Network*
	OutDegree	*InDegree*	*OutDegree*	*InDegree*	*OutDegree*	*InDegree*			
Mean	6.500	6.500	2.703	2.703	3.141	3.141	2.327	0.410	0.242
Std Dev.	7.344	5.019	3.219	2.865	4.527	3.622	4.491	1.087	0.552
Maximum	39	20	22	16	24	18	23.283	5.722	2.816
Minimum	0	0	0	0	0	0	0	0	0
The number of nodes with value 0	3	8	14	15	12	13	26	37	33
The number of nodes with value 0-1(including 1)	–	–	–	–	–	–	18	20	26
The number of nodes with value 1-5 (including 5)	33	21	44	40	46	39	10	5	5

(Table continues on next page)

Table 2.4. Continued

	Degree Centrality†						*Betweenness Centrality‡*		
	Technological Knowledge Network		*Market Knowledge Network*		*Managerial Knowledge Network*		*Technological Knowledge Network*	*Market Knowledge Network*	*Managerial Knowledge Network*
	OutDegree	*InDegree*	*OutDegree*	*InDegree*	*OutDegree*	*InDegree*			
The number of nodes with value 5-10 (including 10)	20	24	4	7	3	9	6	2	0
The number of nodes with value 10-15 (including 15)	3	5	1	1	0	2	2	0	0
The number of nodes with value more than 15	5	6	1	1	3	1	2	0	0

†Nonnormalized Degree Centrality; ‡ Normalized Betweenness Centrality

11 (indegree) nodes in the technological knowledge network which have a degree centrality value greater than 10. The number is pretty small in market (2/2) and managerial knowledge networks (3/3). At the same time, the nodes with higher information control power (with the betweenness centrality value more than 5) in technological knowledge network (10) are more than that in market (2) and managerial knowledge networks (0).

Finally, Table 2.5 presents the structural hole measures of different knowledge networks, which indicates there are difference among the three knowledge networks. When two of the ego's contacts do not share a tie, a structural hole exists between them (Burt, 1992). Some researchers argue that structural holes in a firm's network enhance its knowledge creation (Baum, Calabrese, & Silverman, 2000; McEvily & Zaheer, 1999), and others have found a positive effect of structural holes on individual knowledge creation (Burt, 2004; McFadyen, Semadeni, & Cannella, 2009). Our study finds that there are more structural holes in the technological knowledge network than in market and managerial knowledge networks.

Table 2.5. Structural Hole Measures of Different Knowledge Networks

The Effective Size Statistic	*Technological Knowledge Network*	*Market Knowledge Network*	*Managerial Knowledg Network*
The number of nodes with effective size more than or equal to 3.	29	17	14
The number of nodes with effective size more than or equal to 4.	13	9	10
The number of nodes with effective size more than or equal to 5.	11	6	6

In summary, the social network analysis results indicate technological, market and managerial knowledge flow have different density, transitivity and amounts, and the three knowledge networks represent different network structures. This is demonstrated in the indexes of density, clustering coefficient, network centralization, component, fragmentation and single node structural hole. These findings provide strong support to hypotheses 2a and 2b.

Table 2.6 reports the sources of knowledge ties in the three knowledge networks. In our survey, there are six types of relationship among cooperative firms competitor, supplier, customer, competitor & supplier, competitor & customer, and supplier & customer instead of simple vertical (with competitors) or horizontal relationships (with suppliers and/or customers). Universities are a very important knowledge source with 17.8% of the firms identifying universities as supplying technical knowledge, greater than for competitors (14.0%). This supports the view that research discoveries and techniques made by universities are more accessible than those made by firms, because universities operate largely according to the norms of open science (Owen-Smith & Powell, 2004). However, competitors, suppliers and customers are the biggest source of technological knowledge if we include the categories of the competitor & customer and the competitor & supplier (with the total proportion of 63.8%). Hence, we can conclude that knowledge sources are diverse in a technological knowledge network. Alternatively, suppliers and customers account for a large proportion of the knowledge ties in a market knowledge network (15.7% and 17.6% respectively), and the proportion is larger if we include the source of competitor & customer and competitor & supplier (with the total proportion of 59.8%). Furthermore, universities, suppliers and customers play important roles in providing managerial knowledge (16.5%, 18.6% and 16.5% respectively), and the suppliers and customers become more prominent if we take the contributor of competitor & customer and competitor & supplier into account (with the total proportion of 63%). In our survey, the universities provided firms with managerial knowledge mainly through MBA courses. In addition, the Public Technology Platform in the cluster is an important knowledge contributor in technological knowledge network, and Industry Association in the cluster contributes market and managerial knowledge in respective knowledge networks. Overall, different actors—that is, competitors, suppliers, customers, universities, industry associations and public interest platforms—contribute different amounts of knowledge, and have varying importance among knowledge networks in high technology industrial clusters. This supports hypothesis 3a.

Finally, Table 2.3 presents the results of QAP Correlation test for the three knowledge networks. The findings indicate that a technological knowledge network is significantly correlated with both a market network (r = 0.660) and a managerial knowledge network (r = 0.670), and market knowledge network is significantly correlated with managerial knowledge network (r = 0.584). This supports hypothesis 3b, which proposed that the three kinds of knowledge networks are significantly correlated in high technology industrial clusters. This underpins the perspective that innovation is a complicated and integrated process and requires different sets of knowledge and complementary capabilities.

Table 2.6. The Source of Different Knowledge Ties

Source of Ties	*Technological Knowledge*	*Market Knowledge*	*Managerial Knowledge*
University	17.8%	12.7%	16.5%
Competitor	14.0%	13.7%	9.3%
Supplier	13.4%	15.7%	18.6%
Customer	13.4%	17.6%	16.5%
Competitor & Supplier	11.5%	14.7%	12.4%
Competitor & Customer	11.5%	11.8%	15.5%
Supplier & Customer	2.5%	2.0%	3.1%
Public technology platform	9.6%	2.9%	0
Industry association	6.4%	8.8%	8.2%
Total	100%	100%	100%

DISCUSSION AND CONCLUSION

In this study, we explore the diversity of knowledge content and knowledge flow in three kinds of knowledge networks in regional innovation clusters. This topic has received surprisingly limited attention. Our research represents a start towards integrating different knowledge content into the innovation process and understanding the role of technological, market and managerial knowledge in the value creation equation. What's more, this empirical study has implications for research into the regional knowledge base, regional knowledge networks, regional knowledge source, organizational learning and knowledge-based theory.

Concerning regional knowledge bases, our research focuses on specific knowledge content instead of knowledge properties—such as tacit or codified (Nonaka & Takeuchi, 1995)—or abstract regional knowledge bases—such as analytical, synthetic and symbolic (Asheim & Gertler, 2005; Asheim & Coenen, 2005). Our findings reveal that market and managerial knowledge are each components, along with technological knowledge, of regional knowledge bases. Although extant research suggested the diversity of knowledge to which an organization has access via its partnerships affects its innovation performance, rather than the number of ties per se (Baum et al., 2000), the diversity of knowledge content and different knowledge ties has not been studied. Our empirical results illustrate the presence of three kinds of knowledge, though they have differences in terms of knowledge amount, knowledge density, and transitivity of knowledge flow in the knowledge network. In particular, technological knowledge manifests its prominence in a cluster by the much higher knowledge amount, better transitivity and more active nodes. All of the arrows between firms are bidirectional in technological knowledge network, suggesting the fundamental position of technological knowledge network in regional innovation clusters. This indicates that searching for and acquiring the lacking technological knowledge is the initial mission of an innovation firm. Recognizing different knowledge content in networks is not only helpful to clarifying the concept of knowledge bases, and to extend the field of knowledge management on regional level, but it also helps us understand the interplay between actors in the processes of transmitting and absorbing knowledge to complement their existing knowledge repository.

Regarding regional knowledge networks, though knowledge network research is inherently multilevel and has grown rapidly in the past decades by attracting attention from varying disciplines, it has not been researched from the perspective of different knowledge content. We advance an understanding of the knowledge network beyond technological knowledge network by introducing the concept of market and managerial knowledge networks. The empirical research indicates the three kinds of regional knowledge networks have different structures, heterogeneous knowledge flows, and differentiated knowledge activities of actors, which testifies to the necessity of taking account of different knowledge networks independently. In addition, different knowledge networks also serve as heterogeneous repositories of knowledge in the region, forming the technological knowledge base, market knowledge base and managerial knowledge base of the regional innovation cluster. These new knowledge bases provide an enhanced perspective to interpret the acquiring, transferring, and creating of knowledge (Phelps et al., 2012) in regional innovation clusters.

Taking regional knowledge sources into consideration, previous research took for granted that different collaborators competitors, suppliers, customers, universities, industry associations and public technology platforms provide technological knowledge (Ahuja, 2000; Belderbos et al., 2012; Owen-Smith & Powell, 2004). This is supported in our research. At the same time, our empirical research discovers that the collaborators contribute different knowledge in the knowledge networks. Universities play a very important role in technological and managerial knowledge networks, which is in accordance with the idea that the presence of public research organizations (universities) in a research-intensive industry region should increase information flows and further accelerate innovation by nearby firms (Whittington et al., 2009). Competitors, suppliers and customers, meanwhile, are main knowledge sources in three knowledge networks though differences exist in their specific knowledge contribution. Moreover, industry association and public technology platform also provide knowledge to firms in the cluster. Accordingly, separate knowledge sources provide diverse knowledge content and contribute differentially in providing complementary knowledge to innovative firms in regional innovation clusters.

Concerning organizational learning, the complementarity of different knowledge content is not often mentioned. Rather, prior empirical research has focused on other variables, such as experience accumulation (Nerkar & Roberts, 2004; Pisano, Bohmer, & Edmondson, 2001), the types of social relationships, or the learning outcomes from a network relationship, but the knowledge content in the network is usually technological knowledge. Our study shows that technological knowledge network, market knowledge network and managerial knowledge network coexist, and are significantly correlated in the regional innovation cluster. Similarly, though the growing recognition that the development and deployment of knowledge is a principal source of firm competitive advantage (Eisenhardt & Martin, 2000; Grant, 1996), how to develop and deploy knowledge has not been researched. Our study provides a new perspective to organizational learning and knowledge-based view, showing that combining and integrating different knowledge content is helpful to the process of knowledge creation, transfer, and adoption.

In addition, the study has important managerial implications. In particular, it has shown that technological, market and managerial knowledge exist in networks in the cluster and have close relationships. This indicates the importance of the construction and the accumulation of different types of knowledge rather than merely technological knowledge to the development of enterprises. Simultaneously, our research has discovered the diverse sources of knowledge in different knowledge networks. The findings suggest firms should join networks and come into contact with different actors in the cluster in order to acquire the diverse knowledge they need. In addition, research has argued that alliances can contribute to the development of dynamic capabilities that represent a firm's ability to integrate and (re)combine internal and external knowledge and competences in order to strengthen and renew its competitive advantage (Armstrong & Shimizu, 2007; Eisenhardt & Martin, 2000). We complement this view by emphasizing how firms may integrate and recombine technological, market and managerial knowledge from different cooperators according to their knowledge deficiency instead of just internal and external knowledge.

As with all studies, ours has limitations. A primary one is that we used 56 best firms in the high technology cluster in south China to construct the whole knowledge network. While a contribution of our study is that we use network data from a growing region like the Pearl River Delta in China, using survey data to evaluate the whole network may have limited the depth to which we were able to gain insights about any one dimension of the network. Our findings may not transfer to other settings because of unique characteristics of our sample. Second, prior studies examining the influence of interfirm partnerships on firm innovation provide conflicting results. While many studies have found the more interfirm partners a firm has, the greater its innovation performance (e.g., Ahuja, 2000; Owen-Smith & Powell, 2004), other research suggests an increasing reliance on partners for knowledge can have a diminishing and ultimately negative effect on knowledge creation (Rothaermel & Alexandre, 2009; Wadhwa & Kotha, 2006), suggesting the costs of maintaining an increasing number of interorganizational relationships can exceed their knowledge-creating benefits. We did not empirically test these contradictory opinions in our research. Therefore, future studies could attempt to study different types of clusters rather than just focus on high technology clusters, and compare the knowledge networks between different kinds of clusters. Finally, future research could obtain more data from a cluster, trying to examine how the structure of the knowledge network affects knowledge outcomes, like innovation performance. Continuing to build from the findings of the present study and to address some of these issues will help to build our understanding about knowledge networks and knowledge flows in regional innovation clusters.

ACKNOWLEDGMENTS

The study was supported by the Humanities and Social Sciences Project funded by Chinese Ministry of Education (Project no. 10YJC630315), and Soft Science Project of Guangdong Province funded by Department of Science and Technology of Guangdong Province (Project no. 2012B070200015).

REFERENCES

Abrahamson, E., & Rosenkopf, L. (1997). Social network effects on the extent of innovation diffusion: A computer simulation. *Organization Science, 8*, 289-309.

Agarwal, R., Echambadi, R., Franco, A. M., & Sarkar, M. (2004). Knowledge transfer through inheritance: Spin-out generation, development, and survival. *Academy of Management Journal, 47*, 501-522.

Ahuja, G. (2000). Collaboration networks, structural holes and innovation: A longitudinal study. *Administrative Science Quarterly, 45*, 425-455.

Alcácer, J., & Chung, W. (2007). Location strategies and knowledge spillovers. *Management Science, 53*, 760-776.

Almeida, P., & Kogut, B. (1999). Localization of knowledge and the mobility of engineers in regional networks. *Management Science, 45*(7), 905-917.

Armstrong, C. E., & Shimizu, K. (2007). A review of approaches to empirical research on the resource-based view of the firm. *Journal of Management, 33*, 959-986.

Asheim, B. (1996). Industrial districts as "learning regions": A condition for prosperity. *European Planning Studies 4*, 379-400.

Asheim, B., Boschma, R., & Cooke, P. (2011). Constructing regional advantage: Platform policies based on related variety and differentiated knowledge bases. *Regional Studies, 45*, 893-904.

Asheim, B., & Coenen, L. (2005). Knowledge bases and regional innovation systems: Comparing Nordic clusters. *Research Policy, 34*, 1173-1190.

Asheim, B., Coenen, L., Moodeysson, J., & Vant, J. (2007). Constructing knowledge-based regional advantage: Implications for regional innovation policy. *International Journal of Entrepreneurship and Innovation Management* 7(2-5), 140-155.

Asheim, B., & Gertler, M. (2005). The geography of innovation: Regional innovation systems. In J. Fagerberg, D. Mowery, & R. Nelson (Eds.), *The Oxford handbook of innovation* (pp. 291-317). Oxford, UK: Oxford University Press.

Azoulay, P., Zivin, J. S. G., & Wang, J. L. (2010). Superstar extinction. *Quarterly Journal of Economics, 125*, 549-589.

Bard, J. F., Balachandra, R., & Kaufmann, P. E. (1988). An interactive approach to R&D project selection and termination. *IEEE Transactions on Engineering Management, 35*, 139-146.

Baum, J. A. C., Calabrese, T., & Silverman, B. S. (2000). Don't go it alone: Alliance network composition and startups' performance in Canadian biotechnology. *Strategic Management Journal, 21*, 267-294.

Becattini, G. (1990). The Marshallian industrial district as a socio-economic notion. In P. Pyke, G. Becattini, & W. Sengenberger (Eds), *Industrial districts & inter-firm co-operation in Italy* (pp. 37-51). Geneva, Switzerland: International Institute for Labour Studies (ILO).

Belderbos, R., Carree, M., & Lokshin, B. (2004). Cooperative R&D and firm performance. *Research Policy, 33*, 1477-1492.

Belderbos, R., Gilsing,V., & Lokshin, B. (2012). Persistence of, and interrelation between, horizontal and vertical technology alliances. *Journal of Management, 38*, 1812-1834.

Bell, G. G., & Zaheer, A. (2007). Geography, networks, and knowledge flow. *Organization Science. 18*, 955-972.

Birkinshaw, J., Hamel, G., & Mol, M. J. (2008). Management innovation. *Academy of Management Review. 33*(4), 825-845.

Birkinshaw, J., & Mol, M. (2006). How management innovation happens. *MIT Sloan Management Review, 47*, 81-88.

Bogers, M., Afuah, A., & Bastian, B. (2010). Users as innovators: A review, critique, and future research directions. *Journal of Management, 36*, 857-875.

Bothner, M. S. (2003). Competition and social influence: The diffusion of the sixth-generation processor in the global computer industry. *American Journal of Sociology, 108*, 1175-1210.

Bouty, I. (2000). Interpersonal and interaction influences on informal resource exchanges between R&D researchers across organizational boundaries. *Academy of Management Journal, 43*, 50-65.

Brusco, S. (1990). The idea of the industrial district: Its genesis. In P. Pyke, G. Becattini, & W. Sengenberger (Eds.), *Industrial districts & inter-firm co-operation in Italy* (pp. 10-19). Geneva, Switzerland: International Institute for Labour Studies (ILO).

Burgers, J. H., Van Den Bosch, F. A. J., & Volberda, H. W. (2008). Why new business development projects fail: Coping with the differences of technological and market knowledge. *Long Range Planning, 41*, 55-73.

Burt, R. S. (1992). *Structural holes: The social structure of competition*. Cambridge, MA: Harvard University Press.

Burt, R. S. (2004). Structural holes and good ideas. *American Journal of Sociology, 110*, 349-399.

Buur, J., & Matthews, B. (2008). Participatory innovation. *International Journal of Innovation Management, 12*, 255-273.

Camagni, R. (Ed.) (1991). *Innovation networks: Spatial perspectives*. London, UK: Belhaven.

Chesbrough, H. (2003). *Open innovation*. Boston, MA: Harvard University Press.

Clarysse, B., Wright, M., & Van de Velde, E. (2011). Entrepreneurial origin, technological knowledge and the growth of spin-off companies. *Journal of Management Studies, 40*, 1084-1093.

Cooke, P. (2001). Regional innovation systems, clusters, and the knowledge economy, *Industrial and Corporate Change 10*, 945-974.

Crossan, M. M., & Apaydin, M. (2010). A multi-dimensional framework of organizational innovation: A systematic review of the literature. *Journal of Management Studies, 47*, 1154-1191.

Crossan, M. M., & Hulland, J. (2002). Leveraging knowledge through leadership of organizational learning. In C. Choo & N. Bontis (Eds.), *Strategic management of intellectual capital and organizational knowledge: A collection of readings* (pp. 711-724). New York, NY: Oxford University Press.

Danneels, E. (2002). The dynamics of product innovation and firm competences. *Strategic Management Journal, 23*, 1095–121.

Das, T. K. & Teng, B. (2000). A resource-based theory of strategic alliances. *Journal of Management 26*, 31-61.

Dodgson, M. (1993). *Technological collaboration in Industry: Strategy, policy, and internationalization in innovation*. London, UK: Routledge.

Dyer, J. H., & Nobeoka, K. (2000). Creating and managing a high-performance knowledge-sharing network: The Toyota case. *Strategic Management Journal, 21*, 345-367.

Ebadi, Y. M., & Utterback, J. M. (1984). The effects of communication on technological innovation. *Management Science, 30*, 572-585.

Eisenhardt, K. M., & Martin, J. A. (2000). Dynamic capabilities: What are they? *Strategic Management Journal, 21*, 1105-1121.

Ernst, H., Lichtenthaler, U., & Vogt, C. (2011). The impact of accumulating and reactivating technological experience on R&D alliance performance. *Journal of Management Studies, 48*, 1194-1216.

Faems, D., Van Looy, B., & Debackere, K. (2005). Inter-organizational collaboration and innovation: Towards a portfolio approach. *Journal of Product Innovation Management, 22*, 238-250.

Fleming, L. (2001). Recombinant uncertainty in technological search. *Management Science, 47*, 117-132.

Franke, N., & Shah, S. (2003). How communities support innovative activities: An exploration of assistance and sharing among end-users. *Research Policy, 32*, 157-178.

Freeman, C. (1974). *The economics of industrial innovation*. Harmondsworth, UK: Penguin.

Frohlich, M., & Westbrook, R. (2001). Arcs of integration: An international study of supply chain strategies. *Journal of Operations Management, 19*, 185-200.

Gambardella, A., & Giarratana, M. S. (2010). Organizational attributes and the distribution of rewards in a region: Managerial firms vs. knowledge clusters. *Organization Science. 21*, 573-586.

Gertler, M. S. (1995). Being there: Proximity, organization, and culture in the development and adoption of advanced manufacturing technologies. *Economic Geography, 71*, 1-26.

Goerzen, A., & Beamish, P. W. (2005). The effect of alliance network diversity on multinational enterprise performance. *Strategic Management Journal, 26*, 333-354.

Grant, R. (1996). Toward a knowledge-based theory of the firm. *Strategic Management Journal, 17*(Winter Special Issue), 109-122.

Grimpe, C., & Kaiser, U. (2010). Balancing internal and external knowledge acquisition: The gains and pains from R&D outsourcing. *Journal of Management Studies, 47*, 1483-1509.

Gulati, R. (1998). Alliances and networks. *Strategic Management Journal, 19*, 293-317.

Hagedoorn, J. (2002). Interfirm R&D partnerships: An overview of major trends and patterns since 1960. *Research Policy, 31*, 477-492.

Henderson, R. M., & Clark, K. B. (1990). Architectural innovation: The reconfiguration of existing product technologies and the failure of established firms. *Administrative Science Quarterly, 35*, 9-30.

Hitt, M. A., Bierman, L., Shimizu, K., & Kochhar, R. (2001). Direct and moderating effects of human capital on strategy and performance in professional service firms: A resource-based perspective. *Academy of Management Journal, 44*, 13-28.

Hoskisson, R. E., Hitt, M. A., Wan, W. P., & Yiu, D. (1999). Theory and research in strategic management: Swings of a pendulum. *Journal of Management, 25*, 417-456.

Hunt, S. D. (1999). The strategic imperative and sustainable competitive advantage. *Journal of the Academy of Marketing Science, 27*, 144-159.

Jackson, M. O. (2008). *Social and economic networks*. Princeton, NJ: Princeton University Press.

Jaffe, A. B. (1989). Real effects of academic research. *American Economic Review, 79*, 957-970.

Jaffe, A. B., Trajtenberg, M., & Henderson, R. (1993). Geographic localization of knowledge spillovers as evidenced by patent citations. *Quarterly Journal of Economics. 108*, 577-598.

Kaish, S., & Gilad, B. (1991). Characteristics of opportunities search of entrepreneurs versus executives: Sources, interests, general alertness. *Journal of Business Venturing, 6*, 45-61.

King, N., Anderson, N., & West, M. A. (1992). Organizational innovation: A case study of perceptions and processes. *Work and Stress, 5*, 331-339.

Knoke D., & Yang, S. (2008). *Social network analysis*. Los Angeles, CA: SAGE.

Kogut, B., & Zander, U. (1992). Knowledge of the firm, combinative capabilities, and the replication of technology. *Organization Science, 3*, 383–397.

Kogut, B., & Zander, U. (1996). What do firms do? Coordination, identity and learning. *Organization Science, 7*, 502-518.

Lane, P. J., & Lubatkin, M. (1998). Relative absorptive capacity and interorganizational learning. *Strategic Management Journal, 19*, 461-477.

Laursen, K., & Salter, A. J. (2006). Open for innovation: The role of openness in explaining innovative performance among UK manufacturing firms. *Strategic Management Journal, 27*, 131-150.

Lavie, D. (2007). Alliance portfolios and firm performance: A study of value creation and appropriation in the U.S. software industry. *Strategic Management Journal, 28*, 1187-1212.

Leonard-Barton, D. (1988). Implementation as mutual adaptation of technology and organization. *Research Policy, 17*, 251-267.

Lettl, C., Herstatt, C., & Gemuenden, H. G. (2006). Users' contributions to radical innovation: Evidence from four cases in the field of medical equipment technology. *R&D Management, 36*, 251-272.

Li, T., & Calantone, R. J. (1998). The impact of market knowledge competence on new product advantage: Conceptualization and empirical examination. *Journal of Marketing, 62*, 13-29.

Liebeskind, J. P., Oliver, A. L., Zucker, L., & Brewer, M. (1996). Social networks, learning, and flexibility: Sourcing scientific knowledge in new biotechnology firms. *Organization Science, 7*, 428-443.

Madjar, N., Oldham, G. R., & Pratt, M. G. (2002). There's no place like home? The contributions of work and non-work creativity support to employees' creative performance. *Academy of Management Journal, 45*, 757-767.

Maskell, P., & Malmberg, A. (1999). Localized learning and industrial competitiveness. *Cambridge Journal of Economics, 23*, 167-186.

Marvel, M. R., & Lumpkin, G. T. (2007). Technology entrepreneurs' human capital and its effects on innovation radicalness. *Entrepreneurship Theory and Practice, 31*, 807-828.

McEvily, B., & Zaheer, A. (1999). Bridging ties: A source of firm heterogeneity in competitive capabilities. *Strategic Management Journal, 20*, 1133-1156.

McFadyen, M. A., Semadeni, M., & Cannella, A. A. (2009). Value of strong ties to disconnected others: Examining knowledge creation in biomedicine. *Organization Science, 20*, 552-564.

Mol, M., & Birkinshaw, J. (2009). The sources of management innovation: When firms introduce new management practices. *Journal of Business Research, 62*, 1269-1280.

Morgan, S. L., & Soerensen, A. B. (1999). Parental networks, social closure, and mathematics learning: A test of Coleman's social capital explanation of school effects. *American Sociological Review, 64*, 661-681.

Morrison, E. W. (2002). Newcomers' relationships: The role of social network ties during socialization. *Academy of Management Journal, 45*, 1149-1160.

Mowery, D. C., Oxley, J. E., & Silverman, B. S. (2002). The two faces of partner-specific absorptive capacity: Learning and co-specialization in strategic alliances. In F. J. Contractor & P. Lorange (Eds.), *Cooperative strategies and alliances* (pp. 291-320). Amsterdam, Netherlands: Elsevier Science.

Nambisan, S., & Baron, R. A. (2010). Different roles, different strokes: Organizing virtual customer environments to promote two types of customer contributions. *Organization Science, 21*, 554-572.

Nerkar, A., & Roberts, P. W. (2004). Technological and product-market experience and the success of new product introductions in the pharmaceutical industry. *Strategic Management Journal, 25*, 779-799.

Nonaka, I. & Takeuchi, H. (1995). *The knowledge creating company*. Oxford, UK: Oxford University Press.

Orsenigo, L., Pammolli, F., & Riccaboni, M. (2001). Technological change and network dynamics: Lessons from the pharmaceutical industry. *Research Policy, 30*, 485-508.

Owen-Smith, J., & Powell, W. W. (2004). Knowledge networks as channels and conduits: The effects of spillovers in the Boston biotechnology community. *Organization Science, 15*, 5-21.

Patton, D., & M. Kenney. (2005).The spatial configuration of the entrepreneurial support network for the semiconductor industry. *R&D Management, 35*, 1-16.

Parkhe, A. (1993). Strategic alliance structuring: A game theoretic and transaction cost examination of interfirm cooperation. *Academy of Management Journal, 36*, 794-829.

Phelps, C., Heidl, R., & Wadhwa, A. (2012). Knowledge, networks, and knowledge networks: A review and research agenda. *Journal of Management, 38*, 1115-1166.

Pisano, G. P., Bohmer, R. M. J., & Edmondson, A. C. (2001). Organizational differences in rates of learning: Evidence from the adoption of minimally invasive cardiac surgery. *Management Science, 47*, 752-768.

Prahalad, C. K., & Ramaswamy, V. (2003). *The future of competition: Co-creating unique value with customers*. Boston, MA: Harvard Business School Press.

Porter, M. E. (1990). *Competitive advantage of nations*. New York, NY: Free Press.

Powell, W. (1990). Neither market nor hierarchy: Network forms of organization. *Research in Organizational Behavior, 12*, 295-336.

Powell, W. W., Koput, K.W., & Smith-Doerr, L. (1996). Interorganizational collaboration and the locus of innovation: Networks of learning in biotechnology. *Administrative Science Quarterly, 41*, 116-145.

Powell, W. W., Koput, K. W., Smith-Doerr, L., & Bowie, J. (2002). The spatial clustering of science and capital. *Regional Studies, 36*, 299-313.

Powell, W. W., White, D. R., Koput, K. W., & Owen-Smith, J. (2005). Network dynamics and field evolution: The growth of interorganizational collaboration in the life sciences. *American Journal of Sociology, 110*, 1132-1205.

Priem, R. L., Li, S., & Carr, J. C. (2012). Insights and new directions from demand-side approaches to technology innovation, entrepreneurship, and strategic management research. *Journal of Management, 38*, 346-374.

Reagans, R., & McEvily, B. (2003). Network structure and knowledge transfer: The effects of cohesion and range. *Administrative Science Quarterly, 48*, 240-267.

Rowley, T., Behrens, D., & Krackhardt, D. (2000). Redundant governance structures: An analysis of structural and relational embeddedness in the steel and semiconductor industries. *Strategic Management Journal, 21*(Special Issue), 369-386.

Rosenberg, N. (1982). *Inside the black box: Technology and economics*. Cambridge, UK: Cambridge University Press.

Rosenzweig, E., Roth, A., & Dean, J. (2003). The influence of an integration strategy on competitive capabilities and business performance: An exploratory study of consumer products manufacturers. *Journal of Operations Management, 21*, 437-456.

Rothaermel, F. T., & Alexandre, M. T. (2009). Ambidexterity in technology sourcing: The moderating role of absorptive capacity. *Organization Science, 20*, 759-780.

Sampson, R. C. (2007). R&D alliances and firm performance: The impact of technological diversity and alliance organization on innovation. *Academy of Management Journal, 50*, 364-386.

Saxenian, A. (1996). *Regional advantage: Culture and competition in Silicon Valley and Route 128*. Boston, MA: Harvard University Press.

Schilke, O., & Goerzen, A. (2010). Alliance management capability: An investigation of the construct and its measurement. *Journal of Management, 36*, 1192-1219.

Schilling, M. A., & Phelps, C. C. (2007). Interfirm collaboration networks: The impact of large-scale network structure on firm innovation. *Management Science, 53*, 1113-1126.

Shane, S. (2000). Prior knowledge and the discovery of entrepreneurial opportunities. *Organization Science, 11*, 448-469.

Simonton, D. K. (1999). *Origins of genius: Darwinian perspectives on creativity*. New York, NY: Oxford University Press.

Singh, J. (2005). Collaborative networks as determinants of knowledge diffusion patterns. *Management Science, 51*, 756-770.

Sorenson, O. (2003). Social networks and industrial geography. *Journal of Evolutionary Economics, 13*, 513-527.

Sorenson, O., & Stuart, T.E.. (2008). Bringing the context back in: Settings and the search for syndicate partners in venture capital investment networks. *Administrative Science Quarterly, 53*, 266-294.

Stuart, T. E., Hoang, H., & Hybels, R. C. (1999). Interorganizational endorsements and the performance of entrepreneurial ventures. *Administrative Science Quarterly. 44*, 315-349.

Sullivan, D. M., & Marvel, M. R. (2011). Knowledge acquisition, network reliance, and early-stage technology venture outcomes. *Journal of Management Studies, 48*, 1169-1193.

Teece, D. J. (1992). Competition, cooperation, and innovation: Organizational arrangements for regimes of rapid technological progress. *Journal of Economic Behavior and Organization, 18*, 1-25.

Teece, D. J. (2007). Explicating dynamic capabilities: The nature and microfoundations of (sustainable) enterprise performance. *Strategic Management Journal, 28*, 1319-1350.

Teece, D. J., Pisano, G., & Shuen, A. (1997). Dynamic capabilities and strategic management. *Strategic Management Journal, 18*, 509-533.

TerWal, A. L. J, & Boschma, R. A. (2009). Applying social network analysis in economic geography: Framing some key analytic issues. *Annals of Regional Science, 43*, 739-756.

Tether, B. (2002). Who co-operates for innovation, and why: An empirical analysis. *Research Policy, 31*, 947-967.

Vaccaro, I. G., Jansen, J. J. P., Van Den Bosch, F., & Volberda, H. W. (2012). Management innovation and leadership: The moderating role of organizational size. *Journal of Management Studies, 49*, 28-51.

Vargo, S. L., & Lusch, R. F. (2004). Evolving to a new dominant logic for marketing. *Journal of Marketing, 68*(1), 1-17.

Vassolo, R. S., Anand, J., & Folta, T. B. (2004). Non-additivity in portfolios of exploration activities: A real options based analysis of equity alliances in biotechnology. *Strategic Management Journal, 25*, 1045-1061.

Vickery, S., Droge, J. C., & Calantone, R. (2003). The effects of an integrative supply chain strategy on customer service and financial performance: An analysis of direct versus indirect relationships. *Journal of Operations Management, 21*, 523-539.

von Hippel, E. (1976). The dominant role of users in the scientific instrument innovation process. *Research Policy, 5*, 212-239.

von Hippel, E. (2005). *Democratizing innovation*. Cambridge, MA: MIT Press.

Wadhwa, A., & Kotha, S. (2006). Knowledge creation through external venturing: Evidence from the telecommunications equipment manufacturing industry. *Academy of Management Journal, 49*, 819-835.

West, M. A., & Anderson, N. (1992). Innovation, cultural-values, and the management of change in British hospitals. *Work and Stress, 6*, 293-310.

Whittington, K. B., Owen-Smith, J., & Powell, W. W. (2009). Networks, propinquity, and innovation in knowledge-intensive industries, *Administrative Science Quarterly, 54*, 90-122.

Yli-Renko, H., Autio, E., & Sapienza, H. J. (2001). Social capital, knowledge acquisition, and knowledge exploitation in young, technology-based firms. *Strategic Management Journal, 22*, 587-613.

Zahra, S. A., Ireland, D. R., & Hitt, M. A. (2000). International expansion by new venture firms: International diversity, mode of market entry, technological learning, and performance. *Academy of Management Journal, 43*, 925-950.

CHAPTER 3

STRATEGIC ALLIANCES IN NEW PRODUCT DEVELOPMENT

A Literature Review

Nina Veflen Olsen and Anne Haugen Gausdal

ABSTRACT

Strategic alliances provide important avenues for new product development. As new product development (NPD) is becoming increasingly complex and costly, organizations are more frequently using various partnering arrangements to accomplish their innovative goals. The purpose of this chapter is to contribute to the development of a co-development theory, by presenting a review of existing co-development literature and discussing how strategic alliance affects the NPD process. One of the distinctions we observed in meeting this aim was the difference between studies that investigated alliances in a NPD setting, and those that investigated NPD within alliances. While the first group of studies focused mainly on the alliance formation, the alliance structure and the alliance outcome in a NPD setting, the second group centered around the NPD project formation, NPD structure, and NPD outcome within alliances. Accordingly, a six-category framework for classifying the

Strategic Alliances for Innovation and R&D, pp. 65–83

literature is developed. In addition to conducting an extensive review of recent literature for each category, the theoretical roots are discussed and areas for further research identified. The principal theoretical approach for understanding strategic alliances has been transaction cost economics (TCE). While TCE has been effective in predicting vertical integration among suppliers and buyers in mature industries, we might question if the logic of transaction cost minimization does capture the necessary dynamics for creating new products. This study reveals that other theoretical approaches, such as resource-based view, social network theory and organizational learning have also been utilized to explain these NPD dynamics.

INTRODUCTION

Strategic alliances provide important avenues for new product development. Firms are more and more compelled to leverage their internal strengths with partners' core competencies to enhance or to sustain their capabilities in providing superior products (Gassmann, Enkel, & Chesbrough, 2010; Millson, Ray, & Wilemon, 1996; Mohr & Spekman, 1994). This strong trend toward more research & development (R&D) outsourcing and alliances, that started in large firms in the high-tech industry, concerns nowadays also small and medium-sized firms (SMEs), low-tech industries and the service sector (Gassmann et al., 2011; Hagedoorn & Duysters, 2002). While most new product development (NPD) studies typically focus on activities within a single firm, a new stream of research highlights NPD alliances. Topics such as how NPD alliances are formed and evolve over time, how NPD processes are organized within alliances, and how alliances affect the NPD outcome have been investigated. Yet, despite these studies, more research on NPD alliances is needed. The performance of alliances, in particular, has, received little attention (Gulati, 1998), and in a literature review on co-development alliances Emden, Calantone, and Droge (2006, p. 332) state that there are "only about a dozen of scholarly studies dealing directly with co-development" and "development of an emergent theory addressing key co-development issues is overdue."

The purpose of this chapter is to contribute to the formulation of a co-development theory, by discussing how a strategic alliance in New Product Development affects the NPD process. According to the PDMA (Product Development and Management Association, n.d.), a NPD process is "a disciplined and defined set of tasks and steps that describe the normal means by which a company repetitively converts embryonic ideas into saleable products or services," while a NPD alliance is a "formal arrangement with a separate company for purposes of development" (www.pdma.org/library/glossary.html). Different kinds of NPD alliances exist. An organization might collaborate with customers or lead-users (von

Hippel, 1986), with suppliers (Takeishi, 2001), or other manufacturers (von Hippel, 1987).

This chapter is organized as follows: The next section contains a presentation of the theoretical roots of the alliance literature. Then, the literature review of NPD alliance studies including the development of the new framework is presented, followed by the discussion and concluding remarks.

THEORETICAL ROOTS

The principal theoretical approach for understanding strategic alliances has been transaction cost economics (TCE) (e.g., Hennart, 1988, 1991; Pisano & Teece, 1989; Shan, 1990; Williamson, 1991). TCE predicts that several problems related to governance are handled more efficiently within the borders of a single company. The argument rests upon the assumption that human beings—or at least some of them—are inclined to behave opportunistically. While opportunism in general may be described as self-seeking behavior with guile, different types of such behavior have been identified and described (Wathne & Heide, 2000). Regardless of the specific type of behavior which may arise, TCE assumes that an authority relationship allows the employer to control and correct such behavior. According to this view, the NPD process should be more efficient if it is undertaken in-house.

Alternatively, NPD in a strategic alliance is only superior if the alliance provides resources which are not available for any of the partners acting alone. Such a resource-based view (Das & Teng, 2000a) may explain why a strategic alliance is superior to in-house governance for conducting NPD, even when the assumption of opportunism is maintained. Another alternative view focuses upon the type of governance structure that motivates effort and creativity. It is not evident that an authority relationship (hierarchy) is the most stimulating one in this respect. According to Eisenhardt and Schoonhoven (1996, p. 148), *"Strategic alliance formation is a complex phenomenon involving both strategic and social factors operating within a logic of needs and opportunities for cooperation."* In comparison with transaction-cost thinking, this social network theory (SNT) emphasizes (1) strategic and social factors, not transaction cost, and (2) a theoretical logic of needs and opportunities, not efficiency. The fourth, and last, alternative view is organizational learning (OL), where "Organizations are seen as learning by encoding inferences from history into routines that guide behaviour" (Levitt & March, 1988, p. 319). According to March (1991), a dynamic relation between the exploration of new possibilities and the exploitation of old certainties may result in organizational learning. The sources for learning from direct experience and discovering new possibilities in-house are often limited. The possibility for organizations to learn from the experience and knowledge of their partners may therefore predict alliances to give the best NPD outcome. While the transaction costs that arise for different governance modes must accordingly be balanced by other social, strategic and learning considerations, the choice whether to cooperate or not is not dependent only upon these transaction costs.

While TCE has been effective in predicting vertical integration among suppliers and buyers in mature industries (Hennart, 1991), we might question if the logic of transaction cost minimization captures the necessary dynamics for creating new products. According to Zajac and Olsen (1993), TCE has focused on ex ante and ex post structural properties, without emphasizing the process involved. They suggest that interorganizational exchange relationships over time need to be understood primarily in terms of developmental processes (Zajac & Olsen, 1993). If the TCE framework is unsuitable, we may find that the NPD process is actually more effective in a strategic alliance than when it is conducted in-house, even if TCE predicts the opposite.

A NPD ALLIANCE REVIEW

A fundamental issue within organizational science is related to whether organizations are viewed as things or as processes (Tsoukas & Chia, 2002, Van de Ven & Poole, 2005). This is a distinction that can be traced all the way back to ancient Greek philosophers such as Democritus and Heraclitus (Rescher, 1996): While Democritus pictures *"all of nature as composed of stable material substance or things that changed only in their position in space and time"*, Heraclitus saw reality "*not as a constellation of things, but, one of processes*" (Van de Ven & Poole, 2005). According to Heraclitus, "*Process is fundamental: The river is not an object but an ever-changing flow; the sun is not a thing, but a flaming fire. Everything in nature is a matter of process, of activity, of change*" (Rescher, 1996, p. 10). Around the beginning of the 19th century, philosophers like Peirce, James, Dewey, and Whitehead also highlighted the importance of the process view. They viewed reality as a process and regarded time, change, and creativity as representing the most fundamental facts for understanding the world. Although the process view has not been the most dominant one, an increasing number of studies have lately distinguished and discussed the difference between organizing (as a verb) and organization (as a noun) (Hernes & Weik, 2007; Van de Ven & Poole,

2005; Weick, 1979). The process view has also been applied in studies of alliances. Four ideal-type theories of alliance processes were presented in de Rond and Bouchikhi (2004). These were (1) Life Cycle, (2) Teleology, (3) Evolution, and (4) Dialectic. The life cycle approach describes the life of an alliance from the start to the end, while the teleological approach is typically viewed as a repetitive sequence of goal formulation, implementation, evaluation, and modification of goals based on what was learned or intended by the entity. According to the evolutionary approach, because of the scarcity of resources organizations must compete for survival, and alliance evolution is driven predominantly by context forces. The dialectic approach emphasizes the collision of coexisting but contradictory social forces. By investigating how alliances evolve and by paying attention to the internal tensions between the actors, we would, according to Das and Teng (2000b), gain a better understanding of alliance instability or failure.

Within the new product development (NPD) literature, several process studies have been conducted. Among the most common ones are life cycle studies of intraorganizational development projects, investigating different NPD stages from idea generation to market launch (e.g., Cooper, 1993). Life cycle studies of the diffusion of products from development to market stagnation (e.g., Rogers, 1983) are also very familiar. In addition to NPD life-cycle studies, functional and goal-oriented teleology studies focusing on the NPD outcome are frequently published (e.g., Montoya-Weiss & Calantone, 1994), while evolutionary NPD studies (e.g., Fogelberg & Sanden, 2008; Tschang, 2007) and dialectic studies of innovation (Hellström, 2004) are much rarer.

In our review of recent NPD alliance literature, a framework for classifying the studies emerged. All the studies reviewed belong to either the life cycle type or the teleology type as defined by de Rond and Bouchikhi (2004). One of the distinctions we observed was the difference between studies that investigated alliances in a NPD setting, and those that investigated NPD within alliances. The first group of studies focused mainly on the alliance, while the second group focused on the NPD. Within both of these two groups we found studies focusing on formation, structure and outcome, leaving us with a total of six categories of NPD alliance studies for the literature review: Alliance formation, alliance structure, alliance outcome, NPD project formation, NPD structure, NPD outcome. Table 3.1 shows an outline of the purpose and main question for each category.

Figure 3.1 is developed to illustrate the connection between the six deduced categories of NPD alliance studies and provides an overview of the literature review.

In the following, we will deal with each of the six categories of NPD alliance studies, and we will discuss the questions posed in Table 3.1.

Table 3.1. Categories for Studying NPD Alliance Literature

Category	*Purpose/Telos*	*Main Question*
Alliance formation	To guide the formation of NPD alliances	Why do firms form strategic NPD alliances? How to form NPD alliances?
Alliance structure	Life cycle studies of the NPD alliance, from birth to death	How do alliance structures change over time?
Alliance outcome	To guide the NPD alliance structure	How do NPD alliances perform?
NPD project formation	Life cycle studies of the NPD project, from idea to launch, within an alliance	How does an alliance structure affect the NPD project formation?
NPD structure	Cycle studies of the NPD structure within an alliance	How does an alliance affect the NPD structure over time?
NPD outcome	To guide the NPD process within the alliance	How does an alliance affect the NPD outcome?

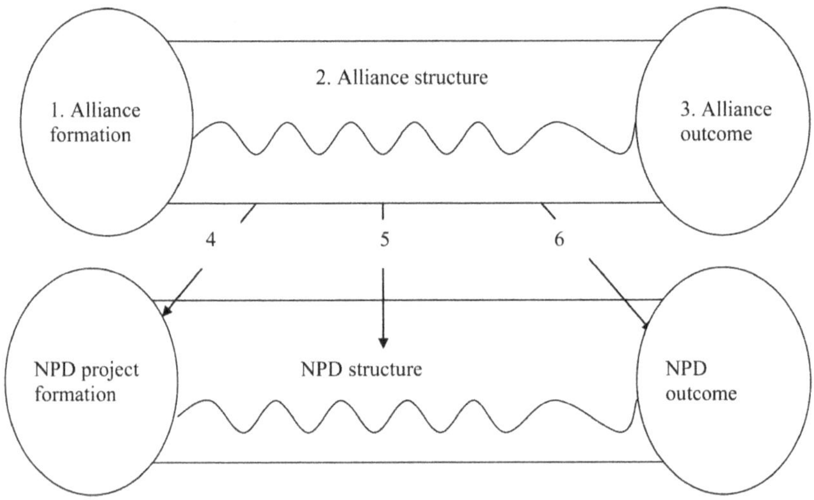

Figure 3.1. Overview of the review.

Alliance Formation

Why do companies form strategic NPD alliances, and how are these alliances formed? Those are the main questions raised within this group of studies. When discussing this topic, scholars base their argumentation on different theoretical roots. Strategic needs and social opportunities are mentioned as two explanations for strategic alliance formation. Leaning on the resource based view of the firm (RBV) as well as on the social network theory (SNT), Eisenhardt and Schoonhoven (1996) state that NPD alliances are formed both when firms are in a vulnerable strategic position and when firms are in strong social positions. In line with the RBV logic, Emden, Calantone, and Droge (2006) argue that strategic partners are chosen on the basis of technical, strategic and relational alignment, while Rothaermel and Boeker (2008) maintain that dyadic complementarities and similarities directly influence alliance formation. Haeussler, Patzelt, and Zahra (2012, p. 218) address the strategic capability perspective by proposing that "new firms can gain significantly from joining alliances by carefully exploiting their own technological capabilities to mitigate the risks involved." Moreover, technological capabilities must be considered when selecting partners. These capabilities are also important in absorbing knowledge from alliances and using it in NPD (Haeussler et al., 2012). Despite the potential benefits, alliances pose significant risks for new high-technology firms (Haeussler et al., 2012). By studying more than four thousand strategic technology alliances in different sectors, Hagedoorn (1993) found that complementary technology and market access constitute the most important motives to engage in partnerships.

According to the SNT logic, alliance formations are affected by different social ties (Gulati, 1995; Tiwana, 2008), as well as by the firm's centrality and structural hole position in network relations (Lin, Yang, & Demirkan, 2007). These strategic capability and social network approaches are quite different from the stand taken by researchers who emphasize the importance of transaction costs. Firms which pursue technology alliances are, according to Robertson and Gatignon (1998), likely to have less commitment to product-specific assets, higher technological uncertainty, higher capability of measuring performance, more successful alliance experience, and compete in lower growth product categories. This is in line with Hagedoorn's (1993) findings of contractual arrangements as the dominating (75%) form in technology partnering, while combination of different arrangements was dominating (64%) the form of market partnering. Motivation to overcome market failure is seen as the reason for forming strategic alliances within the TCE tradition (Williamson, 1985). In sum, RBV, SNT and TCE leave us with three different explanations for alliance formation: (1) strategic need (resources and capabilities), (2) social opportunity (network ties), and (3) market failure avoidance (transaction costs).

Alliance Structure

How does the alliance structure change over time? That is the main question investigated within another group of studies, which mostly lean on theories from organizational learning (OL). How learning processes in alliances mediate between initial conditions and outcome is investigated by Doz (1996), who found alliance projects to be highly evolutionary. Successful alliance projects go through a sequence of interactive cycles of learning, reevaluation and readjustments, while failing projects were characterized by inertia and little learning. The dynamics of exploitation and exploration (March, 1991) in interorganizational NPD projects are investigated by Holmqvist (2004), who proposes a conceptual framework where experiential learning is a driving force behind much interorganizational change in the form of transformation between exploitation and exploration. Another dynamic theory of alliances, proposed by Kumar and Nti (1998), argue that the development path of an alliance is dependent on how the partners assess and react to discrepancies. A conceptual framework for how innovation alliances evolve over time (Linnarson, 2005), as well as an alliance life cycle study (Deeds & Rothaermel, 2003), do also shed light on the alliance structure. Summing up, the literature shows that alliances have a dynamic evolutionary developing structure influenced by the learning and innovation processes.

Alliance Outcome

How do NPD alliances perform? Among the topics investigated are financial returns from alliances (McNamara & Baden-Fuller, 2007), alliances' performance in strengthening alliance network relationships (Kandemir, Yaprak, & Cavusgil, 2006), knowledge exchange performance in alliances (Oxley & Sampson, 2004), more research (Gassmann et al., 2010), and alliance failure due to interfirm rivalry and managerial complexity (Park & Ungson, 2001).

The effects of exploration-exploitation dilemmas on shareholder returns for alliances are explored by McNamara and Baden-Fuller (2007). Superior marketplace performance is, according to Kandemir et al. (2006), a function of how well a firm extracts alliance-related benefits from its partnerships. According to Pilkington (1999), if firms enter a NPD alliance from a position of weakness or without a concerted strategy, they can become reliant on their partner and their design, and manufacturing capabilities may be eroded. Oxley and Sampson (2004) discuss how to maintain sufficiently open knowledge exchange to achieve alliance objectives, while at the same time controlling knowledge flows to avoid unintended leakages of valuable technology. Hess and Rothaermel (2011) argue that, despite managerial challenges, pursuing different activities concurrently can be positive for the innovative outcome in alliances. Completing the knowledge value chain is beneficial, and "pursuing the various stages in the value chain simultaneously provides the basis for unique resource combinations that can be the source of superior performance" (p. 906). Pursuing redundant mechanisms simultaneously results, however, in marginal innovative outcome (Hess & Rothaermel, 2011).

Public research institutions are becoming important partners in NPD strategic alliances. An important outcome of strategic R&D alliances between firms and universities is more open research, which accelerates the race toward commercialization of research results, speeds up the research process itself, energizes the researchers and makes impossible results possible (Gassmann et al., 2010).

Strategic alliances fail, according to Park and Ungson (2001), because of opportunistic hazards as each partner tries to maximize his own individual interests instead of collaborative interests. Also, strategic alliances fail because of the difficulties in coordinating the independent firms. According to the reviewed literature, we can then conclude that alliances influence financial returns, strengthen network relationships, increase knowledge exchange and research. Alliances fail due to interfirm rivalry, opportunistic behavior and managerial coordinating complexity. In addition to the TCE explanations, OL dilemmas as well as RBV topics are emphasized to explain alliance outcome.

NPD Project Formation

How does an alliance affect the NPD project formation? Four viable project organization options are identified by Gerwin and Ferris (2004): In each case they compare the costs and the benefits of the option with respect to the underlying transaction costs, potential for learning, and the ability to contribute to developing a social relations network. How the alliance's organizational form influences partner ability and incentives to share information is also explored (Sampson, 2007). Horizontal alliances have, according to Rindfleisch and Moorman (2001), lower levels of relational embeddedness and higher levels of knowledge redundancy than vertical alliances. But whereas embeddedness enhances both acquisition and utilization of information in alliances, redundancy diminishes information acquisition yet enhances information utilization. Relying on insights from the new product development literature, Faems, Janssens,

Bouwen, and Looy (2006) argue that formal governance mechanisms may hamper exploring new technological opportunities. Further, they highlight the importance of balancing formal and relational governance, while one of the key factors affecting new product development success in alliances seems to be cooperative competency. The ability to build trust, good communication and efficient coordination is an important means of developing interfirm cooperation (Cousins & Lawson, 2007; Deeds & Hill, 1998; Sivadas & Dwyer, 2000; Walter, Lechner, & Kellermanns, 2007). In sum, a new NPD project may be organized according to formality, horizontal or vertical partner composition, transaction costs (TCE), access to new knowledge (OL) and cooperative competence (SNT).

NPD Structure

How does an alliance affect the product development structure over time? A couple of studies highlight this question. Emphasizing the tension between an alliance logic and an innovation logic, Linnarson (2005) identifies a pattern where changes in the alliance structure trigger changes in the innovation. When plotting the changes to the alliance structures on a timeline together with the changes' triggers, he observed that the findings emerged during the exploration of the innovation came to require a changed alliance structure for commercialization. He argues that the pattern observed illustrates a dependency between the innovation and the planned alliance structure, where the dependency could be from innovation to alliance, or from alliance to innovation. These findings are very much in line with Rothaermel and Deeds (2004), who propose and find support for an integrated product development path leading from exploration alliances, via products in development and exploitation alliances, to products on markets. According to the literature on this topic, new technology ventures using an exploration-exploitation strategy in their product development effort tend to have more new products both under development and on the market. Therefore, NPD structure is explained by the theoretical OL approach.

NPD Outcome

How does an alliance affect the NPD outcome? Reasons leading to success and failure of NPD in alliances, as well as different NPD outcomes from alliances, are questions discussed within this group of studies. Both positive and negative effects, along with the imbalance between larger and smaller firm in an alliance, are among the topics highlighted. According to Kalaignanam, Shankar, and Varadarajan (2007), both the larger and the smaller alliance partner experience significant short-term financial gains when involved in NPD alliances. Gil and de la Fé (1999) point out that the weaker partner becomes increasingly dependent on the other partner. Profit-sharing matters in alliance because partners commit more resources when profits are shared proportionally rather than more equally (Amaldoss, Meyer, Raju, & Rapoport, 2000). Fang (2011) found that, in high-technology strategic alliances, NPD process' interdependence and environmental dynamism moderate the effect of knowledge complementarity on the products' innovativeness in a positive way, whereas expropriation risks moderate the effect negatively. Cousins, Lawson, Peterson, and Handfield (2011) maintain that increased breakthrough scanning gives higher technical proficiency at the firm level, and increases knowledge sharing with suppliers. In NPDs, a combination of a firm's technical capabilities and knowledge exchange with suppliers results in better NPD- and financial performance. Furthermore, the largest challenge to achieve such knowledge exchange is to build enough trust between the partners (Cousins et al., 2011).

Fey and Birkinshaw (2005) argue that the use of external contracting has a net negative impact on R&D performance, while alliance partnering has no significant effect. Leaning on RBV and the capability literature, the effect of alliance experience on the NPD outcome is also discussed. Alliance type and alliance experience moderate, according to Rothaermel and Deeds (2006), the relationship between a high-technology venture's R&D alliance and its new product development. Conversely, Hoang and Rothaermel (2005) state that partner-experience has a negative effect on joint project performance. By studying the impact of a firm's internal structure on communication and performance of NPD projects with strategic alliance partners, Badir, Büchel, and Tucci (2009) found that the intensity level and degree of media richness of communication, along with the organizational attributes, vary with the development cycle time (DCT). NPD projects with short DCT require high intensity and high degree of media richness of communication with the partners. Moreover, such projects profit from a decentralized and flatter organizational structure with a moderate formalization level. On the other hand, NPD projects with long DCT require lower intensity and a smaller degree of media richness of communication with the partners. Therefore, they profit from more formalization and from a more centralized organizational structure with several hierarchical levels (Badir et al., 2009). Badir, Büchel, and Tucci (2012) followed up by arguing that the efficient performance of the NPD project is contingent on how well the actual intensity level and the degree of media-richness of communication fit the requirements. Moreover, they claim that these variables, in alliance-based NPD projects, are enabled by varying degrees of team and managerial empowerment, ranging from limited to significant (Badir et al., 2012). We may conclude that, according to the reviewed literature, NPD processes in alliances lead to significant financial gains, access to new knowledge, new resources and sometimes reduced development cycle time. Factors affecting the outcome of the NPD in an alliance are level of cooperation competence, profit-sharing agreements, capability complementarities and power dependency between the partners. NPD processes fail due to lack of trust. The outcome of the NPD process in alliances is therefore explained by theories such as OL, RBV, and SNT.

Table 3.2 gives an overview of studies in the six categories, their main focus and theoretical roots. Quite a few studies have analyzed NPD alliances, but only a minor number of these deals with NPD structures in alliances.

DISCUSSION AND CONCLUDING REMARKS

The overall conclusion from the literature review is that firms form alliances because of strategic needs, social opportunities and as a strategy to avoid market failure. Alliances have a dynamic evolutionary developing structure influenced by learning and innovation processes. They lead to increased financial returns, strengthened network relationships, improved knowledge exchange and more open research. Alliances seem to fail due to interfirm rivalry, opportunistic behaviour and managerial coordinating complexity. The governance structure influence how partners in existing alliances organize a new NPD project according to formality, transaction costs, access to new knowledge, horizontal or vertical composition of partners and cooperative competence. Alliances using an organizational learning strategy in their NPD efforts tend to have more new products under development and on the market. Finally, NPD processes in alliances lead to significant financial gains, access to new knowledge and new resources, and reduced development cycle time. Factors affecting the outcome of the NPD in an alliance are level of cooperation competence, profit-sharing agreements, capability complementarities and power dependency between the partners; when they fail it is due to lack of trust. Therefore, according to RBV, SNT, TCE and OL theories, companies form, design and gain the outcome of strategic NPD alliances and NPD in strategic alliances on the basis of four major theoretical explanations: (1) strategic needs (resources and capabilities), (2) social opportunities (network ties), (3) market failure avoidance (transaction costs), and (4) access to external knowledge (organizational learning).

Because of the evident scarcity of literature, there is definitely a need for more research on NPD structures in alliances, evolutionary NPD studies and dialectic studies of innovation. In order to analyze how alliances

Table 3.2. Review of NPD Alliance Studies

Category	*Main Focus*	*Theoretical Roots*	*Examples of Studies*
Alliance formation (in a NPD setting)	Why do firms form strategic NPD alliances? How to form NPD alliances? The partner selection process. Alliance formation pattern.	Transaction Cost Economics Social Network Theory Resource Based View	1. Emden, Calantone, & Droge (2006) 2. Eisenhardt & Schoonhoven (1996) 3. Gulati (1995) 4. Hagedoorn (1993) 5. Lin, Yang, & Demirkan (2007) 6. Robertson & Gatignon (1998) 7. Rothaermel & Boeker (2008) 8. Tiwann (2008)
Alliance structure (in a NPD setting)	How do alliance structures change over time? The evolution of cooperation in alliances. The development path of alliances. The learning dynamics in alliances	Organizational learning	1. Deeds & Rothaermel (2003) 2. Doz (1996) 3. Holmqvist (2004) 4. Kumar & Nti (1998) 5. Linnarson (2005)
Alliance outcome (in a NPD setting)	How do NPD alliances perform? Alliance performance: Shareholder returns Network performance Knowledge exchange R&D performance Alliance failure Interfirm rivalry Opportunistic behavior Managerial coordinating complexity	Transaction Cost Economics Organizational learning Resource Based View	1. Gassmann, Enkel, & Chesbrough (2010) 2. Haeussler, Patzelt, & Zahra (2012) 3. Hess & Rothaermel (2011) 4. Kandemir, Yaprak, & Cavusgil (2006) 5. McNamara & Baden-Fuller (2007) 6. Park & Ungson (2001) 7. Pilkington (1999) 8. Rothaermel & Deeds (2006) 9. Oxley & Sampson (2004)

(Table continues on next page)

Table 3.2. Review of NPD Alliance Studies (Conti.)

Category	*Main Focus*	*Theoretical Roots*	*Examples of Studies*
NPD project formation in alliances	How does an alliance structure affect the NPD project formation? How to organize and coordinate NPD in alliances? Organizational factors influencing the NPD project in alliances.	Transaction Cost Economics Organizational learning Social Network Theory	1. Cousins & Lawson (2007) 2. Deeds & Hill (1996) 3. Deeds & Hill (1998) 4. Faems, Janssens, Bouwen, & Looy (2006) 5. Gerwin (2004) 6. Gerwin & Ferris (2004) 7. Millson, Ray, & Wilemon (1996) 8. Rindfleisch & Moorman (2001) 9. Sampson (2007) 10. Sivadas & Dwyer (2000) 11. Walter, Lechner, & Kellermanns (2007)
NPD structure in alliances	How does an alliance affect the NPD structure over time? Pattern of alignment in alliance structure and innovation	Organizational learning	1. Linnarson (2005) 2. Rothaermel & Deeds (2004)
NPD outcome in alliances	How does an alliance structure affect the NPD outcome?	Organizational learning Resource Based View	1. Amaldoss, Meyer, Raju, & Rapoport (2000) 2. Badir, Büchel, & Tucci (2009) 3. Badir, Büchel, & Tucci (2012)
	Reasons leading to success and failure of NPD in alliances • Knowledge exchange • Cooperation competence • Profit-sharing • Development cycle time • Trust leve Outcome of NPD alliances: • R&D performance • Financial gains	Social Network Theory	4. Cousins, Lawson, Peterson, & Handfield (2011) 5. Fang (2011) 6. Fey & Birkinshaw (2005) 7. Gil & Fé (1999) 8. Hoang & Rothaermel (2005) 9. Kalaignaman, Shankar, & Varadarajan (2007) 10. Rothaermel (2001) 11. Rothaermel & Deeds (2006)

affect the product development structure and NPD success or failure, alliances should be compared with alternative ways of organizing the process. A comparison could for instance be made between NPD in an alliance and NPD in a single company. Even though both interfirm and intrafirm product development have been investigated, few scholars have compared interfirm processes with intrafirm processes in the same study. Moreover, in spite of all the NPD alliance studies conducted, few of these focus on the micropolitical level. According to Walter, Lechner, and Kellermanns (2007), "*Alliance-related decision making cannot be adequately understood without explicitly considering the micropolitical context in organizations.*" We therefore claim that there is a need for more comparative studies, more studies at the micropolitical level, more studies on NPD structures in alliances, more evolutionary NPD studies and more dialectic studies of innovation.

The purpose of this chapter was to contribute to the formulation of a co-development theory, by discussing how a strategic alliance in new product development affects the NPD process. To fulfill this purpose, we have deduced six categories of NPD alliance studies: alliance formation, alliance structure, alliance outcome, NPD project formation, NPD structure and NPD outcome. We have, moreover, performed an extensive review of recent literature for each category, discussed their theoretical roots and identified areas for further research.

REFERENCES

Amaldoss, W., Meyer, R. J., Raju, J. S., & Rapoport, A. (2000). Collaborating to compete. *Marketing Science, 19*(2), 105-126.

Badir, Y. F., Büchel, B., & Tucci, C. L. (2009). The performance impact of intrafirm organizational design on alliance's NPD projects. *Research Policy, 38*, 1350-1364.

Badir, Y. F., Büchel, B., & Tucci, C. L. (2012). A conceptual framework of the impact of NPD project team leader empowerment on communication and performance: An alliance case context. *International Journal of Project Management, 30*, 914-926.

Cooper, R. G. (1993). *Winning at new products* (2nd ed.). Reading, UK: Addison-Wesley.

Cousins, P. D., & Lawson. B. (2007). The effect of socialization mechanisms and performance measurement on supplier integration in new product development. *British Journal of Management, 18*, 311-326.

Cousins, P. D., Lawson, B., Peterson K. J., & Handfield, R. B. (2011). Breakthrough scanning, supplier knowledge exchange, and new product development performance. *Journal of Product Innovation Management, 28*, 930-942.

Das, T. K., & Teng, B. (2000a). A resource-based theory of strategic alliances. *Journal of Management, 26*, 31-61.

Das, T. K., & Teng, B. (2000b). Instabilities of strategic alliances: An internal tensions perspective. *Organization Science, 11*, 77-101.

Deeds, D. L., & Hill, C. W. L. (1998). An examination of opportunistic action within research alliances: Evidence from the biotechnology industry. *Journal of Business Venturing, 14*, 141-163.

Deeds, D. L, & Rothaermel, F. T. (2003). Honeymoons and liabilities: The relationship between age and performance in research and development alliances. *Journal of Product Innovation Management, 20*, 468-484.

de Rond, M., & Bouchikhi, H. (2004). On the dialectics of strategic alliances, *Organization Science, 15*, 56-69.

Doz, Y. L. (1996). The evolution of cooperation in strategic alliances: Initial conditions or learning processes? *Strategic Management Journal, 17*, 55-83.

Eisenhardt, K. M., & Schoonhoven, C. B. (1996). Resource-based view of strategic alliance formation: Strategic and social effects in entrepreneurial firms. *Organization Studies, 7*, 136-150.

Emden, Z., Calantone, R. J., & Droge, C. (2006). Collaborating for new product development: Selecting the partner with maximum potential to create value. *Journal of Product Innovation Management, 23*, 330-341.

Faems, D., Janssens, M., Bouwen, R., & Looy, B. V. (2006). Governing explorative R&D alliances: Searching for effective strategies. *Management Revue, 17*(1), 9-29.

Fang, E. (2011). The effect of strategic alliance knowledge complementarity on new product innovativeness in China. *Organization Science, 22*, 158-172.

Fey, C. F., & Birkinshaw, J. (2005). External sources of knowledge, governance mode, and R&D performance. *Journal of Management, 31*, 597-621.

Fogelberg, H., & Sanden, B. A. (2008). Understanding reflexive systems of innovation: An analysis of Swedish nanotechnology discourse and organization. *Technology Analysis & Strategic Management, 20*, 65-81.

Gassmann, O., Enkel, E., & Chesbrough, H. (2010). The future of open innovation. *R&D Management, 40*(3), 213-221.

Gerwin, D., & Ferris, J. S. (2004). Organizing new product development projects in strategic alliances. *Organization Science, 15*, 22-37.

Gil, M. J. A., &. de la Fé, P. G. (1999). Strategic alliances, organisational learning and new product development: The cases of Rover and Seat. *R&D Management, 29*(4), 391-404.

Gulati, R. (1995). Social structure and alliance formation patterns: A longitudinal analysis. *Administrative Science Quarterly, 40*, 619-652.

Haeussler, C., Patzelt, H., & Zahra, S.A. (2012). Strategic alliances and product development in high technology new firms: The moderating effect of technological capabilities. *Journal of Business Venturing, 27*, 217-233.

Hagedoorn, J. (1993), Understanding the rationale of strategic technology partnering: Interorganizational modes of cooperation and sectorial differences. *Strategic Management Journal, 14*, 371-385.

Hagedoorn, J., & Duysters, G. (2002). External sources of innovative capabilities: the preferences for strategic alliances or mergers and acquisitions. *Journal of Management Studies, 39*, 167-188.

Hellström, T. (2004). Innovation as social action. *Organization, 11*, 631-649.

Hennart, J. (1988). A transaction costs theory of equity joint ventures. *Strategic Management Journal, 9*, 361-374.

Hennart, J. (1991). The transaction costs theory of joint ventures: An empirical study of Japanese subsidiaries in the United States. *Management Science, 37*, 483-497.

Hernes T., & Weik, E. (2007). Organization as process: Drawing a line between endogenous and exogenous views. *Scandinavian Journal of Management, 23*, 251-264.

Hess, A. M., & Rothaermel, F.T. (2011). When are assets complementary? Star scientists, strategic alliances, and innovation in the pharmaceutical industry. *Strategic Management Journal, 32*, 895-909.

Hoang, H., & Rothaermel, F. T. (2005). The effect of general and partner-specific alliance experience on joint R&D project performance. *Academy of Management Journal, 48*, 332-345.

Holmqvist, M. (2004). Experiential learning processes of exploitation and exploration within and between organizations: An empirical study of product development. *Organization Science, 15*, 70-81.

Kalaignanam, K., Shankar, F. T., & Varadarajan, R. (2007). Asymmetric new product development alliances: Win-win or win-lose partnerships? *Management Science, 53*, 357-374.

Kandemir, D., Yaprak, A., & Cavusgil, S. T. (2006). Alliance orientation: Conceptualization, measurement, and impact on market performance. *Journal of the Academy of Marketing Science, 34*, 324-340.

Kumar, R., & Nti, K. O. (1998). Differential learning and interaction in alliance dynamics: A process and outcome discrepancy model. *Organization Science, 9*, 356-367.

Levitt, B., & March, J. G. (1988). Organizational learning. *Annual Review of Sociology, 14*, 319-340.

Lin, Z. J., Yang, H., & Demirkan, I. (2007). The performance consequences of ambidexterity in strategic alliance formations: Empirical investigation and computational theorizing. *Management Science, 53*, 1645-1658.

Linnarson, H. (2005). Pattern of alignment in alliance structure and innovation. *Technology Analysis & Strategic Management, 17*(2), 161-181.

March, J. G. (1991). Exploration and exploitation in organizational learning. *Organization Science, 2*, 71-87.

McNamara, P. M., & Baden-Fuller, C. (2007). Shareholder returns and the exploration-exploitation dilemma: R&D announcements by biotechnology firms. *Research Policy, 36*, 548-565.

Millson, M. R., Ray, S. P., & Wilemon, D. (1996). Strategic partnering for developing new products. *Research Technology Management, 39*(3), 41-49.

Mohr, J., & Spekman, R. (1994). Characteristics of partnership success: Partnership attributes, communication behavior, and conflict resolution techniques. *Strategic Management Journal, 15*(2), 135-152.

Montoya-Weiss, M. M., & Calantone, R. (1994). Determinants of new product performance: A review and meta-analysis. *Journal of Product Innovation Management, 11*(5), 397-417.

Oxley, J., & Sampson, R. C. (2004). The scope and governance of international R&D alliances. *Strategic Management Journal, 25*, 723-749.

Park, S. H., & Ungson, G.R. (2001). Interfirm rivalry and managerial complexity: A conceptual framework of alliance failure. *Organization Science, 12*, 37-53.

Pilkington, A. (1999). Strategic alliance and dependency in design and manufacturing. *International Journal of Operations and Production Management, 19*, 460-473.

Pisano, G., & Teece, D. (1989). Collaborative arrangements and global technology strategy: Some evidence from the telecommunications equipment industry. *Research on Technology Innovation, Management and Policy, 4*, 227-256.

Product Development and Management Association (n.d.). Retrieved from www.pdma.org/library/glossary.html

Rescher, N. (1996). *Process metaphysics: An introduction to process philosophy*, Albany, NY: State University of New York Press.

Rindfleisch, A., & Moorman, C. (2001). The acquisition and utilization of information in new product alliances: A strength-of-ties perspective. *Journal of Marketing, 65*(April), 1-18.

Robertson, T. S., & Gatignon, H. (1998). Technology development mode: A transaction cost conceptualization. *Strategic Management Journal, 19*, 515-531.

Rogers, E. M. (1983). *Diffusion of innovations*. New York, NY: Free Press.

Rothaermel, F. T., & Boeker, W. (2008). Old technology meets new technology: Complementarities, similarities, and alliance formation. *Strategic Management Journal, 29*, 47-77.

Rothaermel, F. T., & Deeds, D. L. (2004). Exploration and exploitation in alliances in biotechnology: A system of new product development. *Strategic Management Journal, 25*, 201-221.

Rothaermel, F. T., & Deeds, D. L. (2006). Alliance type, alliance experience and alliance management capability in high-technology ventures. *Journal of Business Venturing, 21*, 429-460.

Sampson, R. C. (2007). R&D alliances and firm performance: The impact of technological diversity and alliance organization on innovation. *Academy of Management Journal, 50*, 364-386.

Shan, W. (1990). An empirical analysis of organizational strategies by entrepreneurial high-technology firms. *Strategic Management Journal, 11*, 129-139.

Sivadas, E., & Dwyer, F. R. (2000). An examination of organizational factors influencing new product success in internal and alliance-based processes. *Journal of Marketing, 64*(1), 31-49.

Takeishi, A. (2001). Bridging inter- and intra-firm boundaries: Management of supplier involvement in automobile product development. *Strategic Management Journal, 22*, 403-433.

Tiwana, A. (2008). Do bridging ties complement strong ties? An empirical examination of alliance ambidexterity. *Strategic Management Journal, 29*, 251-272.

Tschang, F. T. (2007). Balancing the tensions between rationalization and creativity in the video games industry. *Organization Science, 18*, 989-1005.

Tsoukas, H., & Chia, R. (2002). On organizational becoming: Rethinking organizational change. *Organization Science, 13*, 567-582.

Van de Ven, A. H., & Poole, M. S. (2005). Alternative approaches for studying organizational change. *Organization Studies, 26*, 1377-1404.

von Hippel, E. (1986). Lead users: A source of novel product concepts. *Management Science, 32*, 791-805.

von Hippel, E. (1987). Cooperation between rivals: Informal know-how trading. *Research Policy, 16*, 291-302.

Walter, J., Lechner, C., & Kellermanns, F. W. (2007). Disentangling alliance management processes: Decision making, politicality, and alliance performance. *Journal of Management Studies, 60*, 698-710.

Wathne, K. H., & Heide, J. B. (2000). Opportunism in interfirm relationships: Forms, outcomes and solutions. *Journal of Marketing, 64*(4), 36-51.

Williamson, O. E. (1970). *Corporate control and business behavior.* Englewood Cliffs, NJ: Prentice-Hall.

Williamson, O. E. (1991). Comparative economic organization: The analysis of discrete structural alternative. *Administrative Science Quarterly, 35*, 91-121.

Weick, K. E. (1979). *The social psychology of organizing* (2nd ed.). New York, NY: Random House.

Zajac, E. J., & Olsen, C. P. (1993). From transaction cost to transaction value analysis: Implications for the study of interorganizational strategies. *Journal of Management Studies, 30*, 131-145.

CHAPTER 4

NEW PRODUCT PERFORMANCE THROUGH CHANNELING IN SUPPLY CHAIN INNOVATION ALLIANCES

The Nexus of Collaboration Intensity, Divergent Communication Schemes, and Alliance Duration

Ricarda B. Bouncken and Robin Pesch

ABSTRACT

The study analyzes how new product development in strategic alliances is affected by collaboration intensity considering the moderating effects of divergent communication schemes and alliance duration. As indicated by proponents of embedded ties, strong collaboration intensity allows a fast exchange of knowledge and further development of ideas, new insights about technologies and markets. The merits of intensified collaboration

Strategic Alliances for Innovation and R&D, pp. 85–111

will increase speed to market and superiority of newly developed products in strategic alliances for innovation. However, we know little so far on the effects of divergent communication schemes between alliance partners. Communication as a process of transferring and encoding messages is not likely to result in the transfer of identical information if the communicating persons have different conceptualizations and different interpretation systems. We assume that divergent communication schemes between alliance partners cause misunderstandings and reduce the benefits of collaboration intensity on new product superiority and speed to market. We also hypothesize that the negative effect of divergent communication schemes is compensated by greater alliance duration because alliance partners develop a better understanding of each other over time.

Results are derived from a survey study of 253 supply chain innovation alliances in the engineering industry using structural equation modeling. Our findings indicate that greater collaboration intensity advances both new product superiority and speed to market. Yet, divergent communication schemes turn the positive effects into negative effects. If firms persevere with the alliance, greater alliance duration can offset the negative effects.

INTRODUCTION

New product development set up to pursue innovation is one pillar of a firm's competitive advantage (Balachandra & Friar, 1997; Griffin, 1997b; Olson, Walker, & Ruekert, 1995). One avenue to improve the quality and speed of new product development (NPD) are innovation alliances within the supply chain environment. A prominent example is the strategic alliance between the South Korean electric supplier LG and the car manufacturer GM for the development of electric cars.

Research on alliances stresses that strategic alliances are a means for the extension of firms' resource base and the access of new technologies (Das & Kumar, 2007; Grant & Baden-Fuller, 2004). Supply chain innovation alliances draw on such enveloped potentials of complementary and extended resources and utilize information transfer up and down the supply chain. In this vein, several empirical studies have shown that the involvement of suppliers in NPD enhances productivity, speed and quality (Clark, 1989; Primo & Amundson, 2002; Ragatz, Handfield, & Petersen, 2002). Roy, Sivakumar, and Wilkinson (2004) evaluate innovations in supply chains as tools for cost reductions and efficiency increases. Supply chain innovation alliances can help to overcome the challenges associated with technological complexity of NPD processes and barriers of foreign market entry (Littler, Leverick, & Bruce, 1995).

However, prior research has only rudimentarily investigated conditions and success of innovation in supply chain alliances allowing greater interaction, collaboration, and information exchange across vertically aligned firms. We argue that communication processes and joint work allow the detection of new technological advances on the upstream side and consumer needs and new business models on the downstream side. Yet, the joint work and communication is based upon a social structure of participants who have to share and process information, coordinate activities, and work coherently if they want to use the potentials of the supply chain alliance. Intensified collaboration by frequent meetings, intensified relationships, and informal interactions across the supply chain stages allows to process information more deeply and to expand the development of new ideas. This can improve the speed of NPD and the superiority of innovations.

At the same time, supply chain innovation alliances have to cope with different targets, behaviors, and beliefs across firms, which are associated with different routines and different role models of the supply chain allies (Chen & Paulraj, 2004; Fawcett, Magnan, & McCarter, 2008; Lajara & Lillo, 2004; Roy et al., 2004). These divergences can hamper communication thereby reducing the advantages of supply chain innovation alliances. Specifically, divergent communication schemes of partners often rooted in the mental sets of managers in different supply chain stages and embedded in different firm cultures can reduce the quality of new products and speed to market by delaying and restricting understanding. Still, different viewpoints within alliances will help to develop novel ideas and further challenge, upgrade, and refine new product concepts.

Intense transfers of knowledge and of technologies across firms through frequent meetings and co-work inspires development processes, but also is disturbed by communication and coordination problems. Firms can overcome this dilemma by a concept that we refer to as channeling. Channeling describes the transfer, merging, and changing of ideas across persons occurring across supply chain stages in innovation alliances.

This chapter investigates the fundamentals around channeling in supply chain innovation alliances by the study of how collaboration intensity influences NPD performance and how it is affected by partners' divergent communication schemes. As firms can learn to cope with others' differences over time, we further investigate if greater alliance duration facilitates interaction and improves new product superiority and speed to market. We structure the chapter as follows: We start with the development of our conceptual model and the corresponding hypotheses. Herein, we first explain supply chain innovation alliances and then develop our concept of channeling. A description of methodology and of empirical results follows. The last sections contain a discussion and conclusion.

THEORY AND HYPOTHESES

Supply Chain Innovation Alliances

Innovations are new products or services, new production process technologies, new structures or administrative systems, or new plans affecting organizational members usually pursued by structured NPD processes (Damanpour, 1991). Trott (2008, p. 15) evaluates NPD processes leading to innovation as "the management of all activities involved in the process of idea generation, technology development, manufacturing and marketing of a new (or improved) product or manufacturing process or equipment." Common understanding is that innovation is a highly structured, knowledge intensive activity (Wang, Yeung, & Zhang, 2011). It requires information about their environmental and endowment conditions, resource deployments, market knowledge, and technological know-how (Gemünden, Heydebreck, & Herden, 1992; Utterback, 1974). Consequently firms have extended their routine work by NPD activities using the potentials of greater information exchange up and down the supply chain (Azadegan, 2011).

Das and Teng (2000, p. 33) describe strategic alliances as "voluntary cooperative interfirm agreements aimed achieving competitive advantage for the partners." In analogy, we define supply chain innovation alliances as interfirm agreements between two or more firms along the supply chain set up for NPD using information exchange and complementary resources, coordination of joint marketing, or shared production facilities. Strategic alliances can take many forms ranging from informal agreements about joint activities to formalized structures of equity-based joint ventures (Das & Kumar, 2007). Similarly, supply chain innovation alliances can range from arm's-length contracts to equity-based joint ventures.

In the past 2 decades, firms have achieved best practice by using supply chain alliances to reduce costs and increase quality across participating firms while improving coordination and information transfer (Barratt, 2004; Fawcett et al., 2008). Typically, the objective of supply chain management is to create and coordinate manufacturing processes seamlessly across the supply chain while being better than competing supply chains (Anderson & Katz, 1998). Firms within supply chains have started to exchange information across partners, establish idea workshops with different firms in the supply chain, test components, change modules and components, use modular designs, and experiment with numerous forms of alliances to improve innovation and to speed up NPD processes.

Alliance research emphasizes that strategic alliances create value through integration of specialized (Grant & Baden-Fuller, 2004), complementary, and supplementary knowledge (Buckley, Glaister, Klijn, & Tan,

2009). Thus, a fundamental mechanism behind the advantages of supply chain alliances for product innovation is the use of suppliers' specialization advantages and their associated complementarities. A second fundamental mechanism is the use of increased upstream and downstream information transfer and coordination advantages that improve NPD. A third fundamental mechanism is the enlarged resource portfolio at hand, since alliances offer access to other firms' resources for creating competitive advantages (Das & Teng, 2000).

The inclusion of partners across the supply chain permits additional adaptations to tangible and intangible resources or appropriate and timely feedback regarding the product design that increase speed and market success. In the environment of short product life-cycles speed to market is highly relevant (Eisenhardt & Tabrizi, 1995; Griffin, 1997a), which defines the time elapsed between the development of an initial product idea and its market launch (McNally, Akdeniz, & Calantone, 2011). Further, NPD performance is reflected by the benefits it creates for customers, especially in relation to competing products (Lee & Colarelli O'Connor, 2003). Therefore, both new product superiority and speed to market are important targets in supply chain innovation alliances.

Channeling

Supply chain innovation alliances formed to generate and commercialize novel products, services, or technologies (Gulati, 1998) use all three fundamental mechanisms. Yet, information transfer or exchange regarding market opportunities, technological advancements, and new business or technology models across organizational boundaries is influenced by different work styles, targets, and behavioral patterns (Hult, Ketchen, & Slater, 2004; Makhija & Ganesh, 1997). These complicate the definition and selection of a product that is to be implemented in the market. They can hamper development, production, supply, and customer services. At the same time, different ideas, targets, limitations, and potentials need to be considered for the selection of a new product that requires actions, operations, and allows financial returns. Thus, different viewpoints of various partners have positive and negative effects.

We develop channeling as a new concept for the mechanism associated with the sharing of information, the stimulation of new ideas, of new technology concepts, and business models across supply chain allies across supply chain stages and organizational boundaries. Channeling envelops the discussion of ideas, the rejection or change of concepts, the overhaul of concepts, and the reflection on the implementation across allies that aim to use their specific knowledge and competences for the development of new products. Channeling appears directly and indirectly via communication and joint work in the NPD process. It requires understanding and consensus, but also covers disagreement, creativity, and constructive discussions. Channeling is based upon relationships across persons and firms. It is a specific outlet of relationships across firms associated with information processing, disagreement, reflections, and the development of new concepts that can stimulate NPD within supply chain alliances.

Prior research has analyzed diverse factors that can be interpreted as antecedents or performance categories of channeling. These are knowledge exchange (Easterby-Smith, Lyles, & Tsang, 2008), resource combination (Grant & Baden-Fuller, 2004), or joint sensemaking within strategic alliances (Cheung, Myers, & Mentzer, 2011). We acknowledge close collaboration and high communication quality as important conditions of channeling (Cheung et al., 2011; Lawson, Petersen, Cousins, & Handfield, 2009). However, little is known about the role of divergent communication schemes between supply chain allies which influence how channeling develops over time. To close this research gap, we examine interrelation intensity, communication divergence, and alliance duration as components of channeling in supply chain innovation alliances. Figure 4.1 shows our conceptual model indicating all the hypotheses.

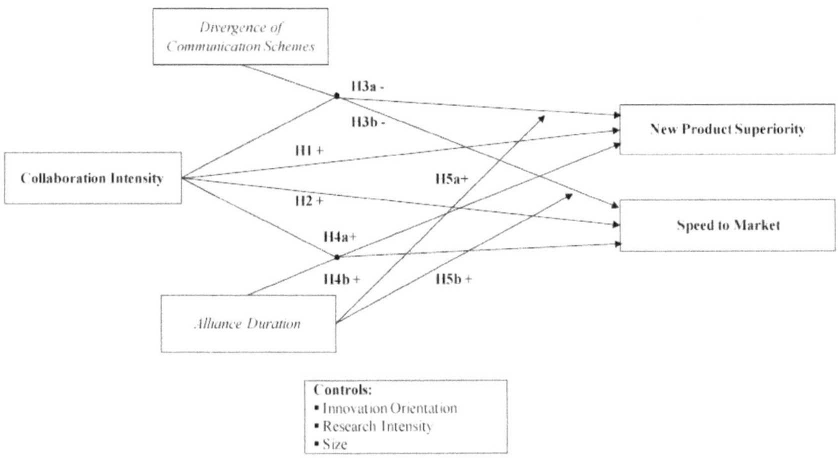

Figure 4.1. Conceptual model.

Interrelation Intensity

High interrelation intensity as "the magnitude of ongoing interactions between partners" shapes relationships, their atmospheres, and, in consequence, determines performance of partnerships (Lin & Germain, 1998, p. 6). Sacco and Vanin (2000, p. 235) refer to interrelation intensity as "the specific channel through which interaction (and specifically cooperation) takes place[,] [...] caus[ing] the emergence of stable social patterns of cooperation through the reinforcement of the interaction between players who [successfully] aspire to the cooperative outcome." Interrelation intensity is often linked to commitment (De Clercq, Sapienza, & Crijns, 2005; Heide & Miner, 1992). Within collaboration across organizational borders, we specify interrelation intensity as collaboration intensity. Consistent with the understanding of collaboration intensity as level of mutual communication and joint engagement (Schleimer & Shulman, 2011), we define collaboration intensity as the strength and frequency of alliance interaction by personal meetings, cultivation of a close relationships, and informal communication that can include knowledge and technical transfers.

Firms need to identify and deploy complementary resources in supply chain innovation alliances based upon understanding, comprehension, and coordination of NPD processes and potentials across partners. This can take advantage from high collaboration intensity in different aspects. Strong interpersonal ties are a channel through which allies learn about each other's competencies (Kale, Singh, & Perlmutter, 2000). Greater collaboration improves firms' capabilities of identifying and evaluating complementary resources which is critical for achieving relational rents (Dyer & Singh, 1998). The greater discussion and understanding across firms allow a better use of the resource portfolio which advances the development of superior new products (Wang et al., 2011). Close and intensive interaction between alliance partners strengthens the understanding of the value and complementarities of alliance partners' resources while improving information exchanges about alliances partners' resources (Lawson et al., 2009). With closer personal interaction of greater collaboration intensity, firms are able to transfer richer information and knowledge than by indirect communication. Interorganizational teams, regular meetings and conferences or social events drive the combination and exchange of resources and knowledge (Autry & Golicic, 2010; Cousins & Menguc, 2006; Lawson et al., 2009; Sobrero & Roberts, 2002; Tether, 2002; Tomlinson, 2010). The importance of collaboration intensity is emphasized for the exchange of implicit knowledge embodied in individual cognition and expertise (Polanyi, 1969) which is a key driver of innovation (Goffin & Koners, 2011; Senker, 1995). The specific characteristics of tacit knowledge require practical experience or personal interaction to be acquired and developed (Nielsen & Nielsen, 2009). Both are rooted in intensified collaboration between alliance partners. Greater collaboration intensity facilitates channeling that additionally shapes a climate of trust (Kale et al., 2000), an important driver of channeling. Lane, Salk, and Lyles (2001) argue that trust improves the ability to understand partners' knowledge. The greater the trust, the more all parties are willing to exchange knowledge and to help each other to understand alliance partners' knowledge. More frequently exchanged information in greater detail and with greater scope from trustful partners can reduce the inherent uncertainty of NPD and improve thereby the quality of product innovation.

In essence, the intensified collaboration stimulates trust and commitment, reduces personal barriers, and provokes openness as persons get to know each other better. Increased trust, commitment, cohesion, and personal understanding further reduce individuals' fear of articulating imprudent ideas and making fools of themselves. It stimulates the free flow of insights, ideas, connotations, meanings, and helps to experiment with new concepts. Greater collaboration intensity as such is a fundament of channeling improving the superiority of product innovation.

> **Hypothesis 1:** *Collaboration intensity between supply chain innovation allies is associated with increased new product superiority.*

Arguments stated above also apply for the improvement of speed to market within alliances across the supply chain. Also referred to as development cycle time, speed to market is defined as the time elapsed between development of the initial product idea and ultimate commercialization (McNally et al., 2011). Yet, further arguments specify the effect of collaboration intensity on speed to market. Ragatz, Handfield, and Scannell (1997) argue that close collaborative NPD offers a faster execution of tasks because differences leading to future problems, such as contradictory requirements, can be identified and solved before problems occur by using an extended resource base. In a similar vein, Zahra, Ireland, and Hitt (2000) postulate that knowledge diversity reduces product development cycles. The improvement of relational capital through close and intensive interaction leads to a development and improvement of interorganizational processes and thereby to the acceleration of NPD (Cousins, Handfield, Lawson, & Petersen, 2006; Nielsen & Nielsen, 2009). By quickening and easing information transfers, collaboration intensity can improve speed to market (Cyril, 2001). Frequent interactions in form of interorganizational teams or regular meetings, and conferences reduce the probability of undesirable developments. A larger body of information, including explicit and tacit knowledge is shared. Frequent interactions improve the speed of information transfers, comprehension, and feedback. They also lead to greater personal attainments that in turn allow a better and faster prediction of partner's behavior. On the downside, high collaboration intensity is time demanding, requires coordination efforts, and on top of this might include cognitive lock-ins (Johnston, McCutcheon, Stuart, & Kerwood, 2004; Littler et al., 1995). The time and efforts of frequent meetings might also distract from the actual work in NPD. Further, the greater information exchange and openness associated with high collaboration intensity opens up windows for opportunism. Still, the speed of the NPD process requires high collaboration intensity to prevent misunderstandings and to use complementary resources and knowledge efficiently.

Hypothesis 2: *Collaboration intensity between supply chain innovation allies is associated with increased speed to market.*

Divergent Communication Schemes

As described before, communication between allies is vital to the success of strategic alliances (Doz & Hamel, 1998) since it helps to ensure understanding and delivers several informational benefits (Butler, 2010). George (2012) emphasizes that communication is needed to coordinate learning processes, to send and acquire knowledge, and to promote interpersonal creativity for creating innovation within alliances. However, firms within supply chain innovation alliances have to cope with divergent communication schemes that relate to dissimilar behavioral patterns, modes of interpretation (Daft & Weick, 1984), and ways of information expression (Lavie, Haunschild, & Khanna, 2012).

A communication scheme is an implicit understanding of how to codify, transfer and interpret information, which also develops in organization and reduces coordination efforts. Differences of interpretation and articulation can root in various factors such as the organizational environment (Daft & Weick, 1984) or the organizational (Lavie et al., 2012) and national culture (Das & Kumar, 2010). Divergent communication schemes become visible in different degrees of directness, formality (Haghirian, 2011; Lavie et al., 2012) and context-orientation of communication (Hall & Hall, 1990). Low context orientation describes a communication process in which the meaning of a message is rooted in the coded, explicit, transmitted information, whereas in high context communication processes the meaning of a message is strongly embedded in the context that surrounds the information (Hall & Hall, 1990).

Divergent communication schemes can induce misunderstanding and frustration, since sender and receiver have different mind sets and expectations on the communication process. Yet, resolving these misunderstandings might stimulate a constructive dialogue, an interchange of different viewpoints, and as such stimulate the development of novel ideas. These can offset cognitive lock-ins and advance product innovation.

However, divergent communication schemes impede channeling by the limitation of the participating firms' capability to assimilate knowledge exchanged in alliances. If individuals misunderstand their alliance partners' knowledge, the organization cannot effectively absorb and apply the obtained knowledge for developing superior products (Lane & Lubatkin, 1998; Malhotra, Gosain, & El Sawy, 2005). A firm's capability to value, assimilate, and apply new knowledge from a particular alliance partner is referred to as partner-specific absorptive capacity (Dyer & Singh, 1998; Lane & Lubatkin, 1998; Malhotra et al., 2005), based on Cohen and Levinthal's (1990) concept of absolute absorptive capacity. Partner-specific absorptive capacity is a critical determinant of knowledge sharing and thereby of achieving relational rents (Dyer & Singh, 1998). We argue that divergent communication schemes in sum impede the positive effects of collaboration intensity on new product superiority.

> **Hypothesis 3a:** *The positive relationship between collaboration intensity and new product superiority is weaker when firms have greater divergent communication schemes. In other words, divergent communications schemes between supply chain innovation allies negatively moderate the relationship between collaboration intensity and new product superiority.*

Similar argumentation holds for the effects on speed to market. Low usage of partner-specific absorptive capacity caused by divergent communication schemes impede channeling and as such the effective combination and exchange of complementary knowledge for new products and optimized interorganizational routines. Divergent communication schemes hamper the positive benefits of close collaboration on speed to market. Misunderstandings caused by dissimilar expectations regarding communication processes are a barrier for the resolution of contradictory requirements, unexpected tasks, and problems. They require additional effort and time. Thus, the positive effect of collaboration intensity on speed to market is reduced by increasingly divergent communication schemes.

> **Hypothesis 3b:** *The positive relationship between collaboration intensity and speed to market is weaker when firms have greater divergent communication schemes. In other words, divergent communication schemes between supply chain innovation allies negatively moderate the relationship between collaboration intensity and speed to market.*

Alliance Duration

Since, alliance conditions are dynamic in nature (Ariño, 1998) and evolve over time (Doz, 1996), we specify our hypotheses by considering the moderating effect of alliance duration also known as link (Kotabe, Martin, & Domoto, 2003) or relationship duration (Squire, Cousins, & Brown, 2009). In the context of supply chain alliances, alliance duration captures the amount of experiences that supply chain allies have in dealing with each other (Kotabe et al., 2003).

The development of innovative products can take advantage of greater experiences across partners across supply chain stages about their environmental and endowment conditions, resource deployments, market knowledge, technological know-how, customers. Greater partner-specific experiences through longer alliance duration will augment advantages from greater collaboration intensity improving NPD processes via enhanced knowledge transfer (Squire et al., 2009) and coordination. Firms achieve a sounder understanding of the value and complementarities of each other's resources over time (Kotabe et al., 2003). We argue that alliance duration influences the relationship between collaboration intensity and NPD performance. As alliance partners operate jointly over an extended period of time, they develop an understanding of peculiarities and barriers that otherwise would inhibit NPD (Cousins et al., 2006). Over time, interorganizational routines emerge and speed up the NPD process (Grunwald & Kieser, 2007). The development and improvement of interorganizational routines is also driven by the development of partner-specific absorptive capacity which enhances the effective combination and exchange of technological and innovation process knowledge between alliances partners.

Hypothesis 4a: *The relationship between collaboration intensity and new product superiority is greater when firms have longer alliance experience. In other words, alliance duration positively moderates the relationship between collaboration intensity and new product superiority in supply chain innovation alliances.*

Hypothesis 4b: *The relationship between collaboration intensity and speed to market is greater when firms have longer alliance experience. In other words, alliance duration positively moderates the relationship between collaboration intensity and speed to market in supply chain innovation alliances.*

Divergent communication schemes that are fraught with misunderstanding foster disagreements and the fear that the other party is acting opportunistically. If firms are unsure about alliance partners' behavior, they will obstruct their partners' access to their resources to impede opportunistic behavior, such as learning races (Khanna, Gulati, & Nohria, 1998). A reduced fear of opportunistic behavior through longer interaction instead decreases the protectiveness of alliance partners and facilitates the combination, exchange, and joint creation of knowledge (Lane et al., 2001; Squire et al., 2009). Ongoing interaction across partners develops relational capital which refers to the level of trust, respect, and friendship between alliance partners (Cousins et al., 2006; Kale et al., 2000; Lawson et al., 2009). It can shape a personal attachment across partners that can help to reduce negative effects due to communicational misunderstanding (Handfield & Bechtel, 2002).

Moreover, modes of interpretations and communication schemes develop over time. They do not only emerge through internal stimuli in the firm. They are also contingent on external factors and communication across organizational borders. As such, supply chain partners can mutually develop more compatible communication schemes or develop tolerance for the divergent communication scheme that they experience in the supply chain alliances. Thus, the development of partner-specific absorptive capacity requires time, since trust and a shared understanding of each other are time demanding determinants of partner-specific absorptive capacity.

We argue that alliance partners learn how to cope with divergent communication schemes or develop joint schemes in the course of the alliance work. Thus, the moderating effect of divergent communication schemes on the relationship between collaboration intensity and new product superiority is influenced by alliance duration. Negative effects through divergent communication schemes that reduce the superiority of new products and the speed to market diminish or can be compensated by longer alliance duration.

Hypothesis 5a: *Negative effects of divergent communication schemes on the relationship between collaboration intensity and new product superiority are positively moderated by alliance duration: greater alliance duration mitigates the negative effects.*

Hypothesis 5b: *Negative effects of divergent communication schemes on the relationship between collaboration intensity and speed to market are positively moderated by alliance duration: greater alliance duration mitigates the negative effects.*

EMPIRICAL STUDY

Questionnaire Design and Sample

The engineering industry is an interesting industry for the analysis of supply chain innovation alliances. This industry has to cope with strong innovative pressures driven by fast technological development across national borders. The engineering industry is mainly characterized by small and medium firms which have limited technological, financial, and personnel resources for innovation activities (Dilk, 2009). Supply chain innovation alliances help to overcome firms' scarcity of resources. Prior to the survey, we carried out two interviews with two executives of the VDMA, Europe's largest engineering association. Both executives stated that the German engineering and technology industry is mainly characterized by small and medium firms which have limited technological, financial, and personnel resources for innovation activities and that supply chain innovation partnerships are widely used to overcome this scarcity of resources. Therefore, we assume that the engineering and technology industry is appropriate for the purpose of our study.

For the survey, we contacted 1,200 executives of firms operating in the engineering industry through a standardized questionnaire in spring 2011. This approach corresponds to the selection of key informants knowledgeable about innovation and alliance matters (Lechner, Dowling, & Welpe, 2006). We received 253 completed questionnaires at a response rate of about 21.08%. Our sample consists of 59 suppliers and 194 manufacturers. To reduce common method bias, we collected data on performance from secondary sources on the firms' research intensity and financial performance. The average sales of the firms in 2010 were approximately 241 million Euros while the average number of workers employed by these organizations amounted to 1,207. The firms' average R&D intensity, defined as the share of R&D expenses to sales, was over 11%.

Scales

Collaboration intensity indicates the strength and frequency of alliance interaction. We measured collaboration intensity by using three items of the scale for technical exchange across units and firms from Kotabe et al. (2003). For new product superiority we adapted the scale of Lee and Colarelli O'Connor (2003). We operationalized speed to market as the time elapsed between development of the initial product idea and commercialization (McNally et al., 2011) by applying three items taken from Rindfleisch and Moorman (2001) as well as from Liao, Fei, and Chen (2007). As communication can be described by the amount of context surrounding information and the degree of directness (Haghirian, 2011; Hall & Hall, 1990), we measured the divergence of communication schemes by asking respondents to compare their (I) context orientation and (II) directness of communication with the context orientation and directness of alliance partner's communication. To measure alliance duration we asked respondents for how many months the firm has been collaborating with its partner. We further included the following control variables: innovation orientation, research intensity and firm size. Innovation orientation is a dynamic capability that assists a firm's proclivity, openness and inclination to generate and distribute novel ideas within alliances (Bouncken, Teichert, & Koch, 2007). It strongly influences the innovation output of firms and therefore should be controlled. We used the following items from Hurley and Hult (1998): the firm's engagement towards (I) creativity in their firm, (II) the constant search for new product ideas, and (III) the participation of personnel in developing novel ideas. Research intensity is defined as research and development investments divided by a firm's sales volume (Cohen & Levinthal, 1990; Hoskisson & Hitt, 1988). This key figure is widely acknowledged as an influential factor on new product superiority and speed to market (Grabowski & Vernon, 1990; Hoskisson & Hitt, 1988; Kamin & Schwartz, 1982). Prior research has shown that the size of a firm affects its adaptability and responsiveness (Hannan & Freeman, 1984) which are of fundamental importance for the fast development of superior new products. In order to control for the impact of firm size, we included each firm's total number of employees in our study.

To avoid common method bias, we applied several procedural steps (Podsakoff, MacKenzie, Lee, & Podsakoff, 2003). First, we reduced the item social desirability effect and evaluation apprehension by assuring all respondents complete confidentiality and anonymity during data collection and analysis. Second, to reduce item ambiguity we avoided vague concepts as well as double barreled questions and kept questions simple (Tourangeou, Rips, & Rasinski, 2000). Besides these procedural controls, we also tested for the presence of common method bias via Harmon's single-factor test, by conducting an unrotated principal component factor analysis on all the variables measured. Five factors with eigenvalues above 1 were identified. The largest did not account for a majority of the variance (37.6%). Therefore, we conclude that the presence of common method bias is unlikely in our study.

Method

To test our hypotheses, we applied structural equation modeling using the Mplus 5.0 software (Muthén & Muthén, 2009). The derived structural equation model reflects the relationships between the latent dependent and independent constructs. Before testing our hypotheses, we applied confirmatory factor analysis (CFA) to validate the measurement model. The goodness-of-fit indices testing the correspondence of the predicted from the proposed model all indicate a very good model fit with CFI = .972, RMSEA .054 and SRMR .040 (Schermelleh-Engel, Moosbrugger, & Müller, 2003). In addition, we carefully examined convergent and discriminant validity. All indicators show good fit (see Table 4.1).

Table 4.1. Construct Measurement

Constructs and Items	Stand. factor loadings	Indicator reliability	α	Composite Reliability	AVE	Fornell-Larcker-Ratio
Collaboration Intensity:			0.90	0.92	0.74	0.24
We regularly meet with our partner.	0.86	0.74				
We cultivate a close relationship to our partner.	0.89	0.80				
There is informal communication with our partner.	0.83	0.69				
New Product Superiority: In the collaboration our innovations…			0.88	0.88	0.71	0.92
incorporate technology that was new to customers.	0.81	0.66				
offer benefits that were new to customers.	0.85	0.72				
introduce many completely new features to the market.	0.87	0.76				
Speed to Market: In the collaboration our innovations			0.86	0.86	0.68	0.96
are developed and launched faster than our typical product development speed.	0.86	0.75				
are developed and launched faster than our major competitor's new products.	0.84	0.71				
are developed and launched faster than the industry norm.	0.77	0.60				
Divergence of Communication Schemes: Compared to us, the collaboration partner has…			0.75	0.77	0.63	0.21
different context use in communication (more task-oriented or associated with more context information).	0.90	0.81				
different communication directness (things are either addressed more directly or more indirect/paraphrased).	0.67	0.45				
Innovation Orientation:			0.76	0.76	0.51	0.81
We strongly encourage creativity in our firm.	0.707	0.50				
We constantly search for new product ideas.	0.703	0.49				
We encourage all personnel in developing novel (product) ideas.	0.734	0.54				

All factor loadings are significant ($t > 3.1$ respectively $p < 0.001$)

Results

Table 4.2 presents the results of our structural models. The first model investigates control variables while the second model tests hypotheses 1 and 2, both supported by the results. The association between collaboration intensity and new product superiority as well as speed to market is positive and significant ($\beta = .292, p < .01$; $\beta = .111, p < .1$).

Table 4.2. Structural Parameters

Paths			Standardized β-Coefficients Model 1	Model 2	β-Coefficients Model 3	Model 4
Controls						
Log(Number of Employees)	→	New Product Superiority	0.022	0.068	0.028	0.040
Log(Number of Employees)	→	Speed to Market	-0.025	-0.028	-0.011	-0.004
R&D Intensity	→	New Product Superiority	0.061	0.098	0.010	0.011*
R&D Intensity	→	Speed to Market	0.102	0.114*	0.009*	0.011**
Innovation Orientation	→	New Product Superiority	0.450***	0.381***	0.546***	0.496***
Innovation Orientation	→	Speed to Market	0.556***	0.526***	0.602***	0.547***
Direct Effects						
Collaboration Intensity	→	New Product Superiority		0.292***	0.203***	0.636*
Collaboration Intensity	→	Speed to Market		0.111*	0.036	0.421
Divergent Communication	→	New Product Superiority			0.142	
Divergent Communication	→	Speed to Market			-0.037	
Log(Alliance Duration)	→	New Product Superiority				0.046
Log(Alliance Duration)	→	Speed to Market				0.025
Interaction Effects						
Collaboration Intensity x Divergent Communication	→	New Product Superiority			-0.187**	
Collaboration Intensity x Divergent Communication	→	Speed to Market			-0.245***	
Collaboration Intensity x Log(Alliance Duration)	→	New Product Superiority				-0.078
Collaboration Intensity x Log(Alliance Duration)	→	Speed to Market				-0.074

Standardized path coefficient significant at *p < 0.1, **p < 0.05; ***p < 0.01

Model 3 examined the moderation of these effects. First, we tested the negative moderation effect of divergent communication schemes by examination of the latent interactions between direct effect and moderator. The association between the interaction term (collaboration intensity X divergent communication schemes) and new product superiority is negative and significant ($\beta = -.187$, $p < .05$). The association between the interaction term and speed to market is negative and significant as well (–.245, $p < .01$). In both cases communication schemes operate as complete moderators as there is no significant association between the moderator and the dependent variables. Our results support hypotheses 3a and 3b. Divergent communication schemes reduce the positive effect of collaboration intensity on new product superiority and speed to market.

To further specify the moderation effects, we plotted the interaction terms (see Figure 4.2). The straight line indicating highly divergent communication schemes has a lower positive incline on new product superiority compared to the straight line indicating low divergent communication schemes. In the case of speed to market, the straight line indicating high divergent communication schemes has even a negative incline compared to the positive incline of the straight line indicating low divergent communication schemes. This explicates the negative moderation by divergent communication schemes.

For the next set of moderation hypotheses, we tested the interaction term between collaboration intensity and alliance duration. The associations between the interaction term (collaboration intensity x alliance duration) and new product superiority ($\beta = -.078$) and between the interaction term and speed to market ($\beta = -.074$) are both insignificant ($p > .1$). Hypotheses 4a and 4b have to be rejected. Alliance duration moderates neither the relationship between collaboration intensity and new product superiority in supply chain innovation alliances nor that between collaboration intensity and speed to market.

The next analyses tested a moderation effect under highly divergent communication schemes that can be understood as a second level or as a two level moderation. We specified the moderation analyses by considering a joint interaction effect of divergent communication schemes and alliance duration. To include both interaction effects in one model, we applied multigroup analysis. In multigroup analysis we tested the direction of the postulated association between collaboration intensity and the dependent variables in case of highly divergent communication schemes. To specify results on supply chain innovation alliances with highly divergent communication schemes, we carried out a median spilt on divergent communication schemes to separate alliances with highly divergent communication schemes. Tests were then carried out on alliances with high divergent communication schemes. To check if alliance duration moderates

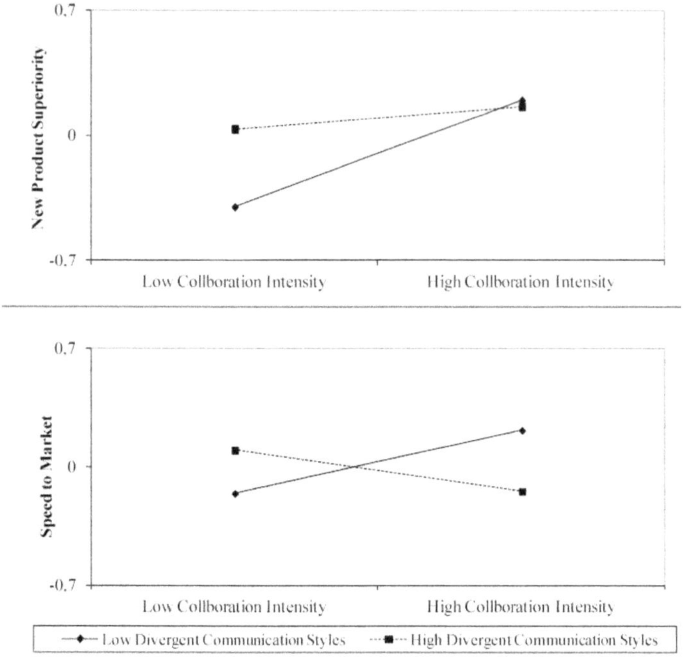

Figure 4.2. Interaction effect of divergent communication schemes.

the relationship between collaboration intensity and performance under highly divergent communication schemes we further had to classify firms into those with short and those with long alliance duration. Subsequently, we tested whether or not the path coefficient between collaboration intensity and new product superiority and between collaboration intensity and speed to market were significantly dissimilar between the two groups of short and long alliance duration when highly divergent communication schemes existed.

Therefore, we spilt the data set into four groups using median splits of the constructs related to the divergence of communication schemes and alliance duration: (I) low divergence of communication schemes and short alliance duration, (II) high divergence of communication schemes and short alliance duration, (III) low divergence of communication schemes and long alliance duration, and (IV) high divergence of communication schemes and long alliance duration.

To test the moderating effect of alliance duration, we solely have to compare the path coefficients (see Table 4.3) between groups (I) and (IV). Here we show the groups (I) and (III) only for completeness. The association between collaboration intensity and new product superiority is significantly negative in the group of high divergence of communication schemes and short alliance duration (II) ($\beta = -.376$, $P < .1$), whereas it is significantly positive in the group of high divergence of communication schemes and long alliance duration (IV) ($\beta = .406$, $p < .1$). From this we conjecture that longer alliance duration does turn the negative effect positive. We checked if the difference is significant by using a chi-square test. For the test of the significance we constrained the hypothesized path to be equal in both groups. Then we compared chi-square values of the constrained and unconstrained model. We discovered a significant chi-square difference ($\Delta\chi^2 = 5.25$ $\Delta df = 1$, $p < .05$). A significant chi-square difference shows that the unconstrained model fits significantly better than the constrained model. Since the path coefficients differ significantly between the short and long duration groups, hypothesis 5a is supported. Negative effects of divergent communication schemes on the relationship between collaboration intensity and product superiority are positively moderated by alliance duration: greater alliance duration turns the negative effects into strongly positive effects.

Table 4.3. Structural Parameters of Multigroup Analysis

Collaboration Intensity → New Product Superiortiy

		Divergence of Communication Schemes	
		low	high
Duration	low	(I) 0.299*	(II) -0.376*
	high	(III) 0.329***	(IV) 0.406*

$\text{II} < \text{IV}: \Delta\chi^2 (\Delta df) = 5.25^{**}(1)$

Collaboration Intensity → Speed to Market

		Divergence of Communication Schemes	
		low	high
Duration	low	(I) 0.242*	(II) -0.377**
	high	(III) -0.054	(IV) 0.073

$\text{II} < \text{IV}: \Delta\chi^2 (\Delta df) = 3.20^{*}(1)$

Significant at $^{*}p < 0.1$; $^{**}p < 0.05$; $^{***}p < 0.01$

We also found support for hypothesis 5b. Estimations show a significant difference ($\Delta\chi^2 = 3.20$ $\Delta df = 1$, $p < .1$) for the relationship between collaboration intensity and speed to market across group (2) and (4) ($\beta = -.377, p < .05$; $.073, p > .1$). Negative effects of divergent communication schemes on the relationship between collaboration intensity and speed to market are positively moderated by alliance duration: greater alliance duration compensates the negative effects, the negative effect becomes nonsignificant.

DISCUSSION

This research was motivated by the increasing practice of supply chain alliances for innovation targets while exchanging information, expertise, and resources across firm boundaries. Prior research has started to analyze NPD in supply chains, but overlooked the social interactions and different behavioral patterns on which we concentrate in this chapter. The starting point of our research is the introduction of the concept of channeling. Allies have to cope with different work styles, targets, behavioral patterns, and in particular with divergent communication schemes, which complicate NPD. Channeling as the discussion of new ideas, the rejection or change of new product concepts, the change of concepts, and the reflection on the implementation across allies helps to overcome these barriers and to find a common line for NPD and implementation across vertically aligned firms. Our results indicate that channeling requires close and intensive interaction, time for developing a shared understanding of each other, and for learning how to cope with divergent communication schemes. Overall, our results contribute to the relational view: we show that supply chain partners can improve their NPD performance in terms of new product superiority and speed to market (Dyer & Hatch, 2006; Dyer & Singh, 1998). Close and intensive interaction between firms facilitates relational capital like trust and partner-specific absorptive capacity, the ability to identify and evaluate alliance partners' complementarities. It opens an avenue for channeling.

We enrich prior research on strategic alliances by showing that divergent communication schemes negatively moderate the relationship between collaboration intensity and NPD performance. Previous research has postulated that communication problems and misinterpretations hamper alliance success (Andersen, 1999; Andersen & Christensen, 2000; Simonin, 1999). However, research has not investigated how the divergence of communication schemes shapes NDP performance. We show that the positive effects of increased collaboration intensity on new product superiority as well as speed to market are negatively affected by divergent communication schemes.

Even though our findings also indicate that greater alliance duration does not advance the positive effect of greater collaboration intensity on NPD, we find that greater alliance duration compensates the negative effects on new product superiority and can rotate the negative effects on speed to market rooted in divergent communication schemes. In this we support and extend prior findings on alliance duration. The empirical study of Squire et al. (2009) showed no significant moderation of alliance duration on the relationship between buyer-supplier cooperation and knowledge transfer. Kotabe et al. (2003) indicate that alliance duration positively moderates the relationship between technology transfer and supplier performance improvement. Our results specify a moderation of alliance duration on the relationship between collaboration intensity and NPD performance in alliances with high divergent communication schemes. Rather, mature and experienced alliances can better cope with divergent communication schemes than juvenile, newer ones. This is in line with the argumentation that alliance partners develop a better understanding of each other and learn how to cope with divergent communication schemes over time. Finally, divergent communication schemes have mainly dysfunctional effects on NPD performance at the beginning of alliances.

In sum we can stress that supply chain alliances require intense interaction, channeling and experience. This can have important implications for managers. First of all, supply chain innovation alliances should be designed for intensive collaboration to trigger channeling. Informal socialization mechanisms such as communication guidelines and social events should be used to overcome the dysfunctional effects rooted in high divergent communication schemes. Such mechanisms help to create a better understanding among partners (Lawson et al., 2009). Socialization mechanisms are necessary at the beginning of alliances because our results show that in this period communication schemes have only dysfunctional effects. The beginning of alliances seems therefore the critical stage which determines alliance success or failure. To avoid alliance failure, alliance managers have to pay special attention to misunderstandings and communication problems at this stage. These implications can also be fruitful for strategic alliances in general as supply chain innovation alliances are a special type of strategic alliances.

However, this study is not without limitations. Our findings rely on data from one participant of the alliance. Beyond that, the focus of our study on a single industry can further limit the generalizability of our findings. A further weakness is the disregard of the innovation phase in which firms ally. We therefore recommend future studies addressing this topic with additional data from both alliance partners to avoid possible distortions and moreover attempt to investigate further industries.

CONCLUSION

The aim of this study was to examine the effects of collaboration intensity on NPD in supply chain alliances using channeling. It revealed that collaboration intensity enhances new product superiority and speed to market in supply chain innovation alliances. Yet, the positive effects of close interaction can be mitigated by communicational differences. Divergent communication schemes negatively moderate these relationships. Yet, the negative effects are reduced and partially compensated by longer alliance duration. As such, firms in supply chain innovation alliances should improve collaboration intensity and give their alliances time when subject to different communication schemes.

REFERENCES

Andersen, P. H. (1999). Organizing international technological collaboration in subcontractor relationships: An investigation of the knowledge-stickiness problem. *Research Policy, 28*, 625-642.

Andersen, P. H., & Christensen, P. R. (2000). Inter-partner learning in global supply chains: Lessons from Novo Nordisk. *European Journal of Purchasing & Supply Management, 6*, 105-116.

Anderson, M. G., & Katz, P. B. (1998). Strategic sourcing. *International Journal of Logistics Management, 9*, 1-13.

Ariño, A., de la Torre, J. (1998). Learning from failure: Towards an evolutionary model of collaborative ventures. *Organization Science, 9*, 306-325.

Autry, C. W., & Golicic, S. L. (2010). Evaluating buyer-supplier relationship-performance spirals: A longitudinal study. *Journal of Operations Management, 28*, 87-100.

Azadegan, A. (2011). Benefiting from supplier operational innovativeness: The influence of supplier evaluations and absorptive capacity. *Journal of Supply Chain Management, 47*(2), 49-64.

Balachandra, R., & Friar, J. H. (1997). Factors for success in R&D projects and new product innovation: A contextual framework. *IEEE Transactions on Engineering Management, 44*, 276-287.

Barratt, M. (2004). Understanding the meaning of collaboration in the supply chain. *Supply Chain Management: An International Journal, 9*, 30-42.

Bouncken, R. B., Teichert, T., & Koch, M. (2007). Innovation strategy explored: Innovation orientation's strategy preconditions and market performance outcomes. *Zeitschrift für Betriebswirtschaft, 2*(Special Issue), 19-43.

Buckley, P. J., Glaister, K. W., Klijn, E., & Tan, H. (2009). Knowledge accession and knowledge acquisition in strategic alliances: The impact of supplementary and complementary dimensions. *British Journal of Management, 20*, 598-609.

Butler, C. J. (2010). Internal and lateral communication in strategic alliance decision making. *Management Decision, 48*, 698-712.

Chen, I. J., & Paulraj, A. (2004). Towards a theory of supply chain management: The constructs and measurements. *Journal of Operations Management, 22*, 119-150.

Cheung, M.-S., Myers, M. B., & Mentzer, J. T. (2011). The value of relational learning in global buyer-supplier exchanges: A dyadic perspective and test of the pie-sharing premise. *Strategic Management Journal, 32*, 1061-1082.

Clark, K. B. (1989). Project scope and project performance: The effect of parts strategy and supplier involvement on product development. Management Science, 35, 1247–1263.

Cohen, W. M., & Levinthal, D. A. (1990). Absorptive capacity: A new perspective on learning and innovation. *Administrative Science Quarterly, 35*, 128-152.

Cousins, P. D., Handfield, R. B., Lawson, B., & Petersen, K. J. (2006). Creating supply chain relational capital: The impact of formal and informal socialization processes. *Journal of Operations Management, 24*, 851-863.

Cousins, P. D., & Menguc, B. (2006). The implications of socialization and integration in supply chain management. *Journal of Operations Management, 24*, 604-620.

Cyril, T. (2001). Interdependencies, trust and information in relationships, alliances and networks. *Accounting, Organizations and Society, 26*, 161-191.

Daft, R. L., & Weick, K. E. (1984). Toward a model of organizations as interpretation systems. *Academy of Management Review, 9*, 284-295.

Damanpour, F. (1991). Organizational innovation: A meta-analysis of effects of determinants and moderators. *Academy of Management Journal 34*, 555-590.

Das, T. K., & Kumar, R. (2007). Learning dynamics in the alliance development process. *Management Decision, 45*, 684-707.

Das, T. K., & Kumar, R. (2010). Interpartner sensemaking in strategic alliances: Managing cultural differences and internal tensions. *Management Decision, 48*, 17-36.

Das, T. K., & Teng, B. (2000). A resource-based theory of strategic alliances. *Journal of Management, 26*, 31-61.

De Clercq, D., Sapienza, H. J., & Crijns, H. (2005). The internationalization of small and medium-sized firms. *Small Business Economics, 24*, 409-419.

Dilk, C. (2009). *Innovationskooperationen und -netzwerke im maschinenbau. Unter besonderer Berücksichtigung von Kooperationskompetenz*. Hamburg, Germany: Verlag Dr. Kovač.

Doz, Y. (1996). The evolution of cooperation in strategic alliances: Initial conditions or learning processes. *Strategic Management Journal, 17*, 55-83.

Doz, Y., & Hamel, G. (1998). *Alliance advantage: The art of creating value through partnering*. Boston, MA: Harvard Business School Press.

Dyer, J. H., & Hatch, N. W. (2006). Relation-specific capabilities and barriers to knowledge transfers: Creating advantage through network relationships. *Strategic Management Journal, 27*, 701-719.

Dyer, J. H., & Singh, H. (1998). The relational view: Cooperative strategy and sources of interorganizational competitive advantage. *Academy of Management Review, 23*, 660-679.

Easterby-Smith, M., Lyles, M. A., & Tsang, E. W. K. (2008). Inter-organizational knowledge transfer: Current themes and future prospects. *Journal of Management Studies, 45*, 677-690.

Eisenhardt, K. M., & Tabrizi, B. N. (1995). Accelerating adaptive processes: Product innovation in the global computer industry. *Administrative Science Quarterly, 40*, 84-110.

Fawcett, S. E., Magnan, G. M., & McCarter, M. W. (2008). Benefits, barriers, and bridges to effective supply chain management. *Supply Chain Management: An International Journal, 13*, 35-48.

Gemünden, H. G., Heydebreck, P., & Herden, R. (1992). Technological interweavement: A means of achieving innovation success. *R&D Management, 22*, 359-376.

George, V. P. (2012). Insular stages of alliances and counterintuitive communication needs. In T. K. Das (Ed.), *Management dynamics in strategic alliances* (pp. 101-116). Charlotte, NC: Information Age Publishing.

Goffin, K., & Koners, U. (2011). Tacit knowledge, lessons learnt, and new product development. *Journal of Product Innovation Management, 28*, 300-318.

Grabowski, H., & Vernon, J. (1990). A new look at the returns and risks to pharmaceutical R&D. *Management Science, 36*, 804-821.

Grant, R. M., & Baden-Fuller, C. (2004). A knowledge accessing theory of strategic alliances. *Journal of Management Studies, 41*, 61-84.

Griffin, A. (1997a). The effect of project and process characteristics on product development cycle time. *Journal of Marketing Research, 34*, 24-35.

Griffin, A. (1997b). PDMA research on new product development practices: Uptrading trends and benchmarking best practices. *Journal of Product Innovation Management, 14*, 429-458.

Grunwald, R., & Kieser, A. (2007). Learning to reduce interorganizational learning: An analysis of architectural product innovation in strategic alliances. *Journal of Product Innovation Management, 24*, 369-391.

Gulati, R. (1998). Alliances and networks. *Strategic Management Journal, 19*, 293-317.

Haghirian, P. (2011). *Multinational and cross-cultural management*. Abingdon, UK: Routledge.

Hall, E., & Hall, M. R. (1990). *Understanding cultural differences*. Yarmouth, UK: Intercultural Press.

Handfield, R. B., & Bechtel, C. (2002). The role of trust and relationship structure in improving supply chain responsiveness. *Industrial Marketing Management, 31*, 367-382.

Hannan, M. T., & Freeman, J. (1984). Structural inertia and organizational change. *American Sociological Review, 49*, 149-164.

Heide, J. B., & Miner, A. S. (1992). The shadow of the future: Effects of anticipated interaction and frequency of contact on buyer-seller cooperation. *Academy of Management Journal, 15*, 265-291.

Hoskisson, R. E., & Hitt, M. A. (1988). Strategic control systems and relative R&D investment in large multiproduct firms. *Strategic Management Journal, 9*, 605-621.

Hult, G. T. M., Jr, Ketchen, D. J., & Slater, S. F. (2004). Information processing, knowledge development, and strategic supply chain performance. *Academy of Management Journal, 47*, 241-253.

Hurley, R. F., & Hult, G. T. M. (1998). Innovation, market orientation, and organizational learning: An integration and empirical examination. *Journal of Marketing, 62*(3), 42-54.

Johnston, D. A., McCutcheon, D. M., Stuart, F. I., & Kerwood, H. (2004). Effects of supplier trust on performance of cooperative supplier relationships. *Journal of Operations Management, 22*, 23-38.

Kale, P., Singh, H., & Perlmutter, H. (2000). Learning and protection of proprietary assets in strategic alliances: Building relational capital. *Strategic Management Journal, 21*, 217-237.

Kamin, M. I., & Schwartz, N. L. (1982). *Market structure and innovation*. Cambridge, UK: Cambridge University Press.

Khanna, T., Gulati, R., & Nohria, N. (1998). The dynamics of learning alliances: Competition, cooperation, and relative scope. *Strategic Management Journal, 19*, 193-210.

Kotabe, M., Martin, X., & Domoto, H. (2003). Gaining from vertical partnerships: Knowledge transfer, relationship duration, and supplier performance improvements in the U.S. and Japanese automotive industries. *Strategic Management Journal, 24*, 293-316.

Lajara, B. M., & Lillo, F. G. (2004). SMEs and supplier alliances use: An empirical analysis. *Supply Chain Management: An International Journal, 9*, 71-85.

Lane, P. J., & Lubatkin, M. (1998). Relative absorptive capacity and interorganizational learning. *Strategic Management Journal, 19*, 461-477.

Lane, P. J., Salk, J. E., & Lyles, M. A. (2001). Absorptive capacity, learning, and performance in international joint ventures. *Strategic Management Journal, 22*, 1139-1161.

Lavie, D., Haunschild, P. R., & Khanna, P. (2012). Organizational differences, relational mechanisms, and alliance performance. *Strategic Management Journal, 33*, 1453-1479.

Lawson, B., Petersen, K. J., Cousins, P. D., & Handfield, R. B. (2009). Knowledge sharing in interorganizational product development teams: The effect of formal and informal socialization mechanisms *Journal of Product Innovation Management, 26*, 156-172.

Lechner, C., Dowling, M., & Welpe, I. (2006). Firm networks and firm development: The role of the relational mix. *Journal of Business Venturing, 21*, 514-540.

Lee, Y., & Colarelli O'Connor, G. (2003). The impact of communication strategy on launching new products: The moderating role of product innovativeness. *Journal of Product Innovation Management, 20*, 4-21.

Liao, S.-h., Fei, W.-C., & Chen, C.-C. (2007). Knowledge sharing, absorptive capacity, and innovation capability: An empirical study of Taiwan's knowledge intensive industries. *Journal of Information Science, 33*, 340-359.

Lin, X., & Germain, R. (1998). Sustaining satisfactory joint venture relationships: The role of conflict resolution strategy. *Journal of International Business Studies, 29*, 79-198.

Littler, D., Leverick, F., & Bruce, M. (1995). Factors affecting the process of collaborative product development: A study of UK manufacturers of information and communications technology products. *Journal of Product Innovation Management, 12*, 16-32.

Makhija, M. V., & Ganesh, U. (1997). The relationship between control and partner learning in learning-related joint ventures. *Organization Science, 8*, 508-527.

Malhotra, A., Gosain, S., & El Sawy, O. A. (2005). Absorptive capacity configurations in supply chains: Gearing for partner-enabled market knowledge creation. *MIS Quarterly, 29*, 145-187.

McNally, R. C., Akdeniz, M. B., & Calantone, R. J. (2011). New product development processes and new product profitability: Exploring the mediating role of speed to market and product quality. *Journal of Product Innovation Management, 28*, 63-77.

Muthén, L. K., & Muthén, B. O. (2009). *Mplus. Statistical analysis with latent variables*. Los Angeles, CA: Muthén & Muthén.

Nielsen, B. B., & Nielsen, S. (2009). Learning and innovation in international strategic alliances: An empirical test of the role of trust and tacitness. *Journal of Management Studies, 46*, 1031-1056.

Olson, E. M., Walker, O. C., & Ruekert, R. W. (1995). Organizing for effective new product development: The moderating role of product innovativeness. *Journal of Marketing, 59*(1), 48-62.

Podsakoff, P. M., MacKenzie, S. B., Lee, J.-Y., & Podsakoff, N. P. (2003). Common method biases in behavioral research: A critical review of the literature and recommended remedies. *Journal of Applied Psychology, 88*, 879-903.

Polanyi, M. (1969). *Personal knowledge. Towards a post-critical philosophy*. Chicago, IL: University of Chicago Press.

Primo, M. A. M., & Amundson, S. D. (2002). An exploratory study of the effects of supplier relationships on new product development outcomes. *Journal of Operations Management*, 20, 33-52.

Ragatz, G. L., Handfield, R. B., & Petersen, K. J. (2002). Benefits associated with supplier integration into new product development under conditions of technology uncertainty. *Journal of Business Research, 55*, 389-400.

Ragatz, G. L., Handfield, R. B., & Scannell, T. V. (1997). Success factors for integrating suppliers into new product development. *Journal of Product Innovation Management, 14*, 190-202.

Rindfleisch, A., & Moorman, C. (2001). The acquisition and utilization of information in new product alliances: A strength-of-ties perspective. *Journal of Marketing, 65*(2), 1-18.

Roy, S., Sivakumar, K., & Wilkinson, I. F. (2004). Innovation generation in supply chain relationships: A conceptual model and research propositions. *Journal of the Academy of Marketing Science, 32*, 61-79.

Sacco, P. L., & Vanin, P. (2000). Network interaction with material and relational goods: An exploratory simulation. *Annals of Public and Cooperative Economics 71*, 229-259.

Schermelleh-Engel, K., Moosbrugger, H., & Müller, H. (2003). Evaluating the fit of structural equation models: Tests of significance and descriptive goodness of-fit measueres. *Methods of Psychological Research Online, 8*, 23-74.

Schleimer, S. C., & Shulman, A. D. (2011). A comparison of new service versus new product development: Configurations of collaborative intensity as predictors of performance. *Journal of Product Innovation Management, 28*, 521-535.

Senker, J. (1995). Tacit knowledge and models of innovation. *Industrial and Corporate Change, 4*, 425-427.

Simonin, B. L. (1999). Ambiguity and the process of knowledge transfer in strategic alliances. *Strategic Management Journal, 20*, 595-623.

Sobrero, M., & Roberts, E. B. (2002). Strategic management of supplier-manufacturer relations in new product development. *Research Policy, 31*, 159-182.

Squire, B., Cousins, P. D., & Brown, S. (2009). Cooperation and knowledge transfer within buyer–supplier relationships: The moderating properties of trust, relationship duration and supplier performance. *British Journal of Management, 20*, 461-477.

Tether, B. S. (2002). Who co-operates for innovation, and why: An empirical analysis. *Research Policy, 31*, 947-967.

Tomlinson, P. R. (2010). Co-operative ties and innovation: Some new evidence for UK manufacturing. *Research Policy, 39*, 762-775.

Tourangeou, R., Rips, L. J., & Rasinski, K. (2000). *The psychology of survey responses*. Cambridge, UK: Cambridge University Press.

Trott, P. (2008). *Innovation management and new product development* Vol. 4. Harlow, UK: Prentice Hall.

Utterback, J. (1974). Innovation in industry and diffusion of technology. *Science, 183*, 620-626.

Wang, L., Yeung, J. H. Y., & Zhang, M. (2011). The impact of trust and contract on innovation performance: The moderating role of environmental uncertainty. *International Journal of Production Economics, 134*, 114-122.

Zahra, S. A., Ireland, R. D., & Hitt, M. A. (2000). International expansion by new venture firms: International diversity, mode of market entry, technological learning, and performance. *Academy of Management Journal, 43*, 925-950.

CHAPTER 5

INCOMPLETE CONTRACTING IN STRATEGIC ALLIANCES

The Quest For Innovation

Regien Sumo, Geert Duysters, Wendy van der Valk, and Arjan van Weele

ABSTRACT

Though incomplete contracts have generally been viewed as inefficient governance mechanisms in strategic alliances, recent contributions suggest that incomplete contracts are conducive to innovation. By integrating transaction cost economics (TCE) and agency theory (AT) we shed light on this contradiction by stating what determines the degree of contractual completeness and how a less complete (i.e., incomplete) contract affects innovation. By linking these concepts we hope to take a first step towards enhancing our understanding as to how incomplete contracts could affect positive relationship outcomes, as extant literature mainly focuses on the negative effects of incomplete contracts on relationship outcome. According to TCE and AT, a contract is dependent on certain transaction characteristics (e.g., transaction specific investment and behavioral uncertainty). We argue that the higher the degree of these transaction characteristics, the more detailed (i.e. complete) a contract becomes. Furthermore, we state that a

Strategic Alliances for Innovation and R&D, pp. 113–132

less complete (i.e., incomplete) contract enhances a partner's autonomy, which in turn, as shown by existing literature, fosters innovation. It is important to note, however, that the relationship between contractual incompleteness and innovation follows an inverted U-shape: too much contractual incompleteness inhibits innovation, as it may lead partners to display opportunistic behavior. However, linking rewards to performance attenuates the negative relationship between contractual incompleteness and innovation when there is a high degree of contractual incompleteness.

INTRODUCTION

Increased competitive pressure has led firms to focus on their core competences and to outsource noncore activities to external partners (Prahalad & Hamel, 1990). As a result, external partners have become an important source for organizations to improve their innovation performance (Chesbrough, 2003; Chesbrough, Vanhaverbeke, & West, 2006; Huston & Sakkab, 2006). Since partners can be a source of innovation, interorganizational relationships (IOR) have become an important strategy for organizations to complement and supplement their internal innovation strategies (Faems, Janssens, Madhok, & Van Looy, 2008; Faems, Van Looy, & Debackere, 2005; Hagedoorn, 2002). More specifically, IORs are generally found to enhance innovation performance (Faems et al., 2005; Goes & Park, 1997; Teece, Pisano, & Shuen, 1997). However, IORs might also negatively affect innovation because of opportunistic behavior (Malhotra & Lumineau, 2011) or coordination failures that impede the efforts of even well-intentioned parties (Gulati, Lawrence, & Puranam, 2005).

In response, organizations use governance mechanisms such as contracts to mitigate these hazards (Lumineau & Malhotra, 2011; Ring & Van de Ven, 1992). Transaction cost economics (TCE) and agency theory (AT) argue that organizations seeking to minimize hazards and maximize transaction gains should write contracts that are as complete as possible. However, in practice, organizations are simply unable to foresee all future events. Furthermore, they have problems in considering all ex ante and ex post transaction costs. Therefore, contracts are often incomplete (Mayer & Argyres, 2004).

From a transaction cost and agency theory perspective these incomplete contracts basically are poorly designed governance mechanisms, since they do not sufficiently address the transaction characteristics that may result in opportunistic behavior (Goldberg, 1976, 1985; Williamson, 1985). Nevertheless, incomplete contracts offer significant benefits over complete contracts. First, they are more *flexible* by allowing for contingency adaptability; that is, allowing the partner to deal with unforeseen circumstances

(Bernheim & Whinston, 1998; Luo, 2002). Second, they are characterized by more *freedom* regarding the specific details of transactions (Bernheim & Whinston, 1998; Luo, 2002). Thus, incomplete contracts are by definition less prescribing, which in turn might foster innovation (Kim, Cohen, & Netessine, 2007; Ng & Nudurupati, 2010; Ng, Maull, & Yip, 2009). Innovation in this respect is defined as discrete, proactive undertakings that are self-initiated by the partner to advance new or improved ways of delivering the transaction in IORs (Johnson & Medcof, 2007; Wang, Yeung, & Zhang, 2011) this definition of innovation is used throughout the chapter.

The tension between the "inefficiency" of incomplete contracts versus their potential positive impact on innovation is the central issue discussed in this chapter. In this chapter, *we contribute to this debate by focusing on what determines contractual (in)completeness and how the degree of (in)completeness affects innovation in IORs.*

Our reasoning regarding the determinants of contractual completeness and the effects of less complete contracts on innovation is based on arguments from agency and transaction cost theory of contractual governance. We will build on the work of many authors in this field (e.g., Argyres, Bercovitz, & Mayer, 2007; Argyres & Mayer, 2007; Eisenhardt, 1989; Luo, 2002; Reuer & Arino, 2007; Saussier, 2000; Williamson, 1985). The works, however, build on either transaction cost theory *or* agency theory as none of these authors has yet attempted to combine the two theories to understand the effects of (incomplete) contracts on innovation. This is counterintuitive as the two theories provide different solutions for fostering partner innovation in IORs. As such, we need to consider these theories collectively rather than separately.

This chapter is structured as follows; the next section elaborates on innovation in IORs. Next, we provide an overview of how contractual completeness is dependent on certain transaction characteristics depicted by TCE and AT. Furthermore, building on the literatures on (formal) governance and innovation, we explain how contractual (in)completeness affects innovation. The chapter concludes with implications and avenues for future research concerning formal governance and innovation in IORs.

INNOVATION IN INTERORGANIZATIONAL RELATIONSHIPS

Innovation (both radical and incremental) in products, processes, and services is seen as crucial for the long-term survival of a firm (Faems et al., 2005). Organizations can engage in innovation through internal innovation activities, such as internal development, R&D and corporate entrepreneurship. However, closed innovation practices, in which innovation is initiated based on the firm's own resources and technologies, have been shown to suffer from limitations (Chesbrough et al., 2006). Being focused on their own innovative capabilities limits the firm in responding to changing environments. Internally, firms do not have all the necessary resources to succeed in complex environments. Moreover, firms face difficulties when trying to capture the value of their resources; as a result, the full value potential often remain underexploited (Chesbrough, 2003). Furthermore, internal innovation practices, like R&D, do not fit the current needs of organizations as they are characterized by high costs and risks, slow time-to-market and inflexibility (Vanhaverbeke, Van de Vrande, & Chesbrough, 2008).

Several authors have suggested semistructured and ambidextrous organizational structures as a way to overcome the aforementioned problems associated with internal innovation (Faems et al., 2005). Within those structures, different strategies, that is, exploiting existing business *and* engaging in innovative activities, can coexist (Christensen & Overdorf, 2000; Faems et al., 2005). As a result, external partners have become a critical source of innovative solutions, ideas and technologies (van Echtelt, Wynstra, van Weele, & Duysters, 2008; Roy, Sivakumar, & Wilkinson, 2004), and interorganizational relationships (IORs), such as strategic alliances (Phelps, 2010), have become important supplements to an organization's internal innovation activities (Deeds & Rothaermel, 2003; Faems et al., 2005; Hagedoorn, 2002). Moreover, IORs are thought to enhance (Faems et al., 2005; Goes & Park, 1997; Teece, Pisano, & Shuen, 1997) or even drive innovation (Hamel, 1991; Leonard-Barton, 1995). Organizations may engage in IORs with a variety of external partners such as potential lead users, universities and research centers, suppliers, and even competitors (Faems et al., 2005).

The reasons that IORs positively affect innovation are diverse (Faems et al., 2005). First, through their collaboration, partners have access to complementary resources that are unavailable or difficult to develop in-house, and that may lead to innovation and nonresident expertise (Dooley & O'Sullivan, 2007; Dyer & Singh, 1998; Goes & Park, 1997; Hardy, Phillips, & Lawrence, 2003; King, Covin, & Hegarty, 2003; Sampson, 2004). The complementarity of such resources is especially obvious in vertical collaborations in which the partners are active in different stages of the value chain, and therefore hold different resources (King et al., 2003). Second, collaborations lead to the transfer of complementary knowledge (Dooley & O'Sullivan, 2007; Dyer & Singh, 1998; Goes & Park, 1997; Hardy et al., 2003) which can be used to exploit existing knowledge and explore new opportunities (Im & Rai, 2008). Collaborations may function as channels for transferring and creating knowledge and they incentivize both partners to reveal information and share technology (Goes & Park, 1997). Collaboration provides organizations with opportunities for mutual learning, which stimulates knowledge creation and enhances innovation capabilities (Tsai, 2001). This results in the creation of complementary resources and development of capabilities that would otherwise be difficult to achieve (Faems et al., 2005). As Goes and Park (1997, p. 681) put it: "learning and competencies can be encapsulated in these exchanges, enabling innovative capability to cross more actively between related firms or network partners."

However, not all IORs are equally successful. Estimates rather show that 60 percent of all alliances fail (Faems et al., 2005). Frequently cited reasons for alliance failure are unintended knowledge spillovers, learning races between the parties, diverging opinions on intended benefits, and lack of adaptability and flexibility (Faems et al., 2005). Furthermore, alliances may suffer from opportunistic behavior and coordination failures (Malhotra & Lumineau, 2011). Organizations attempt to mitigate these conflicts and risks by means of governance mechanisms (Lumineau & Malhotra, 2011). One such mechanism to govern IORs is by means of a contract (Williamson, 1985). Contracts can be an important determinant of whether or not the partner engages in innovation (Johnson & Medcof, 2007; Wang et al., 2011). A wide variety of contracts exist; but not all contracts allow the partner to engage in innovation. For example, some contracts are designed in such a way that they permit or incentivize the partner to perform only those activities that are stipulated in the contract (Johnson & Medcof, 2007). In such instances, the partner is hindered from engaging in innovation. Mintzberg (1994) even stated that "formalization is a double edged sword, easily reaching the point where help becomes hindrance" (p. 386).

TCE and AT have extensively been used in research into contract design and its effects on performance (Argyres et al., 2007; Argyres & Mayer, 2007; Eisenhardt, 1989; Luo, 2002; Reuer & Arino, 2007; Saussier, 2000; Williamson, 1985). In the following sections we will integrate these theories to explain how (incomplete) contracts affect innovation. We will first, however, discuss what these two theories consider to be the core components of the degree of contractual (in)completeness.

TRANSACTION COSTS AND CONTRACTING

Transaction Cost Economics

Transaction cost theory has emerged as an important paradigm for understanding how organizations design their interorganizational governance arrangements (Mayer & Argyres, 2004; Poppo & Zenger, 2002). The literature takes its starting point with Coase's (1937) statement that the boundary of a firm is defined by cost-minimizing choices, specifically, whether firms choose to make or buy products and services (as in Anderson & Dekker, 2005). According to Williamson (1979), incomplete information and information asymmetry prevent organizations from writing complete contracts. These information problems are seen as a source of transaction costs that motivate organizations to vertically integrate rather than buy the product or service. More recently, however, transaction cost theory has been used in the emergence of hybrid organizational forms (e.g., alliances, preferred supplier arrangements, and joint ventures) that combine the advantages of buying from the market with the control and oversight of vertical integration (Anderson & Dekker, 2005; Williamson, 1985). Thus, if vertical integration is neither feasible nor economical and if market transactions cannot meet demands for customization, coordination, and collaboration, TCE opts for a middle ground of hybrid organizational forms (Anderson & Dekker, 2005; Williamson, 1985).

In order to ensure successful hybrid organizational forms, organizations make use of formal controls, such as contracts, to govern the IOR (Das & Teng, 2001). Contracts, which provide legally bound, institutional frameworks, specify the obligations of the parties, such as roles, responsibilities, performance expectations, monitoring clauses, and dispute resolution procedures (Barthelemy & Quelin, 2006; Luo, 2002; Zhou & Poppo, 2010). As such, contracts reduce the uncertainty faced by organizational decision makers and the risks stemming from opportunism on the part of one or more contracting parties by controlling the behavior of both parties (Luo, 2002; Poppo & Zenger, 2002; Williamson, 1985; Zhou & Poppo, 2010). In addition, contracts serve as a method for coordinating the transaction by mitigating the risk that misunderstandings will disrupt the collaboration among the parties (Malhotra & Lumineau, 2011; Zhou & Poppo, 2010). According to TCE, the degree of contractual completeness is dependent on three important transaction characteristics, namely: (1) transaction specific investments, (2) frequency of the transaction, and (3) environmental uncertainty (Anderson & Dekker, 2005; Chen & Bharadwaj, 2009; Williamson, 1979). These transaction characteristics potentially trigger increased transaction costs, which undermines the efficiency of economic exchange and determines how complete the contract is (Zhou & Poppo, 2010).

Transaction specific investments identify the degree to which the assets in a given exchange are more valuable in the current exchange than in their next best use (Chen & Bharadwaj, 2009; Leiblein & Miller, 2003). It is the exposure of the transacting parties to ex-post opportunistic holdup caused by the specific investment in these assets (Anderson & Dekker, 2005). Examples are investing in very specific machines, customized technology or training of personnel (Chen & Bharadwaj, 2009; Stump & Heide, 1995). In order to curb the opportunism caused by the holdup hazard, organizations try to design contracts as complete as possible by for, example, implementing close monitoring processes, intellectual property control provisions, and dispute resolutions (Brousseau & Coeurderoy, 2005). Consistent with the above, Reuer and Arino (2007) find that contracts facing transaction specific investments contain detailed pretermination, arbitration and lawsuit clauses.

Frequency relates to the intensity of trade between the two partners. In TCE, frequency is intended to capture the degree to which future transaction opportunities serve as a bond against opportunities in the present period (Anderson & Dekker, 2005). Though prior research states that a certain frequency will provide some degree of mutual trust and therefore reduce the need to draft complete contracts (Chen & Bharadwaj, 2009; Dyer & Singh, 1998; Ring & Van de Ven, 1992), recent theoretical and empirical work into contract design states that parties become familiar with each other's business processes, decision making styles and corporate cultures, and thus improve the ability to draft more detailed contracts (Chen & Bharadwaj, 2009; Mayer & Argyres, 2004; Poppo & Zenger, 2002). Thus, a higher frequency of trade leads to less transaction costs as a result of an increased ability to describe tasks, associated monitoring requirements, and dispute resolutions, thereby increasing the ability to write detailed contracts.

Finally, *environmental uncertainty* occurs when parties involved in a transaction are unable to fully predict and understand the environment in which the transaction takes place (Heide, 1994). It refers to unanticipated and unpredictable changes in circumstances surrounding an exchange (Zhou & Poppo, 2010). Environmental uncertainty will increase the use of explicit clauses to facilitate adjustments as events occur (Masten, 1993). In addition, in case of environmental uncertainty, organizations explicitly formalize roles and processes (Verbeke & Greidanus, 2009; Zhou & Poppo, 2010). Thus, the parties want to seek control and design more complete contracts if environmental uncertainty is more strongly present. Similarly, Barthelemy and Quelin (2006) find that when there is more environmental uncertainty, the more detailed the contract becomes, especially regarding clauses which foster adaptation of the exchange given certain levels of uncertainty (Zhou & Poppo, 2010).

In sum, TCE assumes that the characteristics of the exchange in IORs affect the potential for increased transaction costs and ex-post problems. In such cases, the higher the degree to which these exchange hazards are present, the more TCE opts for more complete contracts (Williamson, 1979).

Agency Theory

Agency theory concerns the well-known agency relationship, in which one party (the principal) delegates work to another party (the agent), which performs that work (Eisenhardt, 1989). AT is first concerned with resolving the agency problem arising from: (a) goal conflict between the agent and the principal, and (b) the difficulty of and cost associated with verifying agent behavior. Second, AT tries to resolve the problem of risk sharing that arises when the principal and the agent have differing risk attitudes (Eisenhardt, 1989). AT determines the most efficient contract (i.e., behavior- or outcome-based contract) with reference to several exchange characteristics that may create ex-post problems such as opportunism (Eisenhardt, 1989; Jensen & Meckling, 1976) and hence influence the degree of contractual completeness. These exchange characteristics are: (1) measurement problems; (2) task complexity; and (3) behavioral uncertainty (Anderson & Dekker, 2005; Chen & Bharadwaj, 2009; Eisenhardt, 1989; Jensen & Meckling, 1976; Mooi & Ghosh, 2010).

Measurement problems refer to difficulties associated with measuring the outcomes of the transaction (Anderson, 1985; Eisenhardt, 1985). When performance cannot be easily verified, it becomes more difficult to assess whether the agent adheres to the contract (Stump & Heide, 1996). This increases the hazard the principal is faced with as the agent might engage in inappropriate and opportunistic behavior. In addition, measurement problems will increase the renegotiation costs that are incurred when further detailing the contract, which adds to transaction costs (Mooi & Ghosh, 2010). As such, organizations design contracts as complete as possible in order to economize on transaction costs by, for example, implementing monitoring mechanisms and dispute resolution clauses (Anderson, 1985; Eisenhardt, 1989; Mooi & Ghosh, 2010).

Task complexity concerns the complexity of the transaction. Complexity is common when both parties contribute complex components or services that interact in unpredictable ways to produce the desired final product or service (Anderson & Dekker, 2005). Task complexity increases the hazards the principal is faced with by, for example, introducing uncertainty about a failure and increasing implementation problems, thereby increasing transaction costs. As such, when task complexity is present, the greater the effort to draft detailed contracts by, for example, implementing detailed clauses concerning monitoring, dispute resolution, and legal obligations (Mooi & Ghosh, 2010).

Finally, *behavioral uncertainty* occurs when parties cannot effectively monitor the performance and behavior of the other party (Zhou & Poppo, 2010). When behavior is difficult to measure, the agent is incentivized to limit its efforts because the principal cannot accurately measure productivity

(Zhou & Poppo, 2010). To mitigate this uncertainty, principals will seek to draft more explicit contracts regarding, e.g., non-performance, roles and responsibility and monitoring clauses (Eisenhardt, 1989; Zhou & Poppo, 2010).

In sum, according to AT, certain transaction characteristics are present in all principal-agent relationships. These characteristics affect the degree of hazard the principal is faced with, thereby increasing transaction costs. As a result, AT suggests that to mitigate the hazards, more complete contracts should be drafted.

Overall, in line with existing research, transaction hazards depicted by TCE or AT are associated with more complete contracts. In the following section we will outline how the degree of contractual (in)completeness affects innovation by integrating these two perspectives.

CONTRACTS AND INNOVATION

Complete contracts aim to cover all potential events that relate to the fulfillment of the contract, thereby assuming that all future events can be foreseen (Saussier, 2000). In reality, however, contracts are always incomplete, at least to a certain extent (Al-Najjar, 1995; Hart & Moore, 1988; Kloyer & Scholderer, 2012; Mayer & Argyres, 2004). Contracts are inevitably incomplete because they, first, relax the rationality assumptions underlying complete contracts. Incomplete contract proponents realize that both parties are subject to bounded rationality, which prevents them from specifying all relevant contractual clauses (Aghion & Holden, 2011; Tirole, 1999). Hence, parties may not identify some contingencies, or not acknowledge the need to specify certain dimensions of contractual performance (Bernheim & Whinston, 1998). Second, organizations balance the ex-ante costs (e.g., negotiation costs) of designing complete contracts with the ex-post costs (e.g., renegotiation costs) of less complete contracts (Crocker & Masten, 1991).

Measured against the complete contract benchmark, an incomplete contract is one that does not take into account all the relevant contractual terms (Saussier, 2000). This incompleteness relates to two dimensions (Bernheim & Whinston, 1998; Luo, 2002). First, incomplete contracts do not make actions sensitive to verifiable events than would appear optimal (Bernheim & Whinston, 1998), meaning that contracts do not include *contingency adaptability*, and are thus characterized by flexibility (Luo, 2002). Second, incomplete contracts do not specify observable obligations and actions for the parties, thereby allowing more freedom for the partner (Bernheim & Whinston, 1998). The degree to which observable obligations and actions are specified in the contract is also known as "*term specificity*"

(Luo, 2002). Whereas term specificity concerns how specific and detailed the clauses are, contingency adaptability involves how to contractually respond to future problems, conflicts, and contingencies (Luo, 2002). This freedom and flexibility can be favorable for innovation as it allows the partner to make decisions at its own discretion and influences the way the transaction is delivered. In the following section we will describe how the degree of incompleteness of a contract affects innovation.

The Effects of Contractual Incompleteness on Innovation

When contracts are incomplete, and hence are faced with less rules and obligations, the focus is more on the outcome rather than on the process of achieving that outcome. As a result, partners have a certain degree of freedom to conduct their work in the way they think is best (Johnson & Medcof, 2007; Wang et al., 2011). An incomplete contract allows the partner more freedom in decision making regarding the delivery of the transaction. The partner can, thus, choose which activities to engage in and which resources to use at their own discretion. As such, for contracts that are characterized by more incompleteness, partners have a higher degree of autonomy. To ensure that partners engage in innovative activities, they should not be constrained by contractual rules and obligations (Wang et al., 2011). Thus, by granting partners with autonomy, the degree to which they can shape and influence their activities increases and they can quickly respond to changes. This autonomy therefore allows the partner the freedom and flexibility to initiate innovative activities (Abbey & Dickson, 1983; Arad, Hanson, & Schneider, 1997; Morris, Kuratko, & Covin, 2007; Paolillo & Brown, 1978).

As noted above, all contracts are to a certain extent incomplete. Within the range of incomplete contracts however, there is further variation: the precise degree to which a contract is incomplete may vary across the range of incomplete contracts. As such, contractual incompleteness may be high, but also very high. In line with TCE and AT, a very high degree of contractual incompleteness also paves the way for opportunistic behavior (e.g., to seek their personal interest at the expense of the collaborative interest) (Eisenhardt, 1989; Shimizu, 2012; Williamson, 1985) as it results in too much autonomy for the partner. When autonomy is very high, even the most reliable partners might be tempted to act opportunistically (Shimizu, 2012). Opportunistic behavior in the context of incomplete contracts include, for example, preparation of own competitive activities, or selling the generated knowledge to a competitor (Kloyer & Scholderer, 2012). We therefore conclude that a very high degree of autonomy harms the overall quality and performance of innovative activities (Shimizu, 2012). Thus, as contracts become very incomplete, transaction costs are likely to rise disproportionally. Therefore, there is a certain inflection point, at which adding further incompleteness proves to be negative.

In sum, although incompleteness of the contract is important to make the partner engage in innovation as it provides them with autonomy to decide how to deliver the transaction, too much contractual incompleteness creates a context that stimulates the partner to act opportunistically by focusing on their own individual objectives, thereby negatively affecting innovation (Guth & MacMillan, 1986; Shimizu, 2012). Similarly, innovation literature suggests that both very high and very low control (thus very high and very low contractual incompleteness) result in negative innovation performance (Shimizu, 2012; Wang et al., 2011).

The Moderating Role of Incentive Schemes

The issue of opportunism in view of contractual incompleteness is essentially a principal-agent problem in which the goals are misaligned (Eisenhardt, 1989). Curbing this opportunism by adding more control and coordination mechanisms is not considered to be a solution (Shimizu, 2012), as such measures will stifle innovation. To illustrate, by incorporating control and coordination mechanisms the contract will explicitly prescribe roles and obligations, determine the content of the transaction, and specify rules for violating contractual agreements (Argyres & Mayer, 2007; Poppo & Zenger, 2002). This obviously hinders innovation by setting the stage for a more rigid relationship (Arad et al., 1997; Wang et al., 2011). Furthermore, extensive control and coordination mechanisms may hamper information exchange between the parties due to clear specifications of what is and is not allowed (Wang et al., 2011). Finally, increased control- and coordination mechanisms can be considered as a sign of distrust (Ghoshal & Moran, 1996), resulting indirectly in a barrier to knowledge transfer.. We therefore argue that a balance between autonomy and appropriate control- and coordination mechanisms is crucial for a successful relationship (Shimizu, 2012).

As an alternative approach to curbing opportunism, AT proposes to use compensation systems in which the partner is paid based on the performance he delivers (Devers, Cannella, Reilly, & Yoder, 2007; Eisenhardt, 1989; Makri, Lane, & Gomez-Mejia, 2006; Shimizu, 2012). In such cases, the rewards of the partner are linked to its performance through incentive schemes that specify performance goals (Argyres & Mayer, 2007; Lyons, 1996). In case rewards are only linked to specific behavior or the use of certain resources, partners tend to engage only in activities that are explicitly rewarded and not in others (such as innovation) (Deckop, Mangel, & Cirka, 1999; Eisenhardt, 1989). In the most extreme case, any new initiative would even be a breach of contract (Johnson & Medcof, 2007). On the other hand, in case rewards are clearly linked to overall performance, partners have more incentives to engage in unspecified activities that allow them to gain a higher performance level. In this case, innovative behavior is stimulated. As such, when partners are paid based on their performance, they will exhibit activities that are in favor of reaching the contracted performance; they will find new and improved ways of delivering the performance to maximize their returns. As such, linking rewards to performance will direct the behavior of the partner towards collaborative goals even when he is faced with contractual incompleteness and has the possibility to behave opportunistically (Shimizu, 2012). There is a growing body of research that provides evidence for a positive relationship between performance incentives and innovation (Abbey & Dickson, 1983; Johnson & Medcof, 2007; Shepherd & DeTienne, 2005; Shimizu, 2012). This clearly calls for more attention for rewarding partners for performance in order to increase innovation and decrease opportunistic behavior.

In sum, Figure 5.1 shows our conceptual model, which considers the relationship between incomplete contracts and innovation. We argue that the higher the degree of transaction characteristics depicted from TCE and AT, the more complete the contract is. However, a detailed contract inhibits the partner to engage in innovative activities due to the restrictions imposed through detailed contractual clauses. As such, we state that a less complete (i.e., incomplete) contract provides the partner with a certain degree of autonomy, which positively affects innovation. It is important to note however that a high degree of contractual incompleteness inhibits innovation, as it may lead partners to display opportunistic behavior. This effect can be mitigated by paying the partner based on their performance. Pay-for-performance can, thus, be viewed as a moderator to the incompleteness-innovation relationship.

DISCUSSION AND CONCLUSIONS

Despite the abundance of literature that addresses the dangers of incomplete contracting, we have argued for the positive effects of the use of incomplete contracts, more specifically on innovation. In contrast to prior studies, which usually are grounded in either of theoretical perspective, we collectively use agency and transaction cost theory of contractual governance to understand the effects of (incomplete) contracts on innovation. We argue that the exchange characteristics specified by these theories, such as transaction specific investments, frequency,

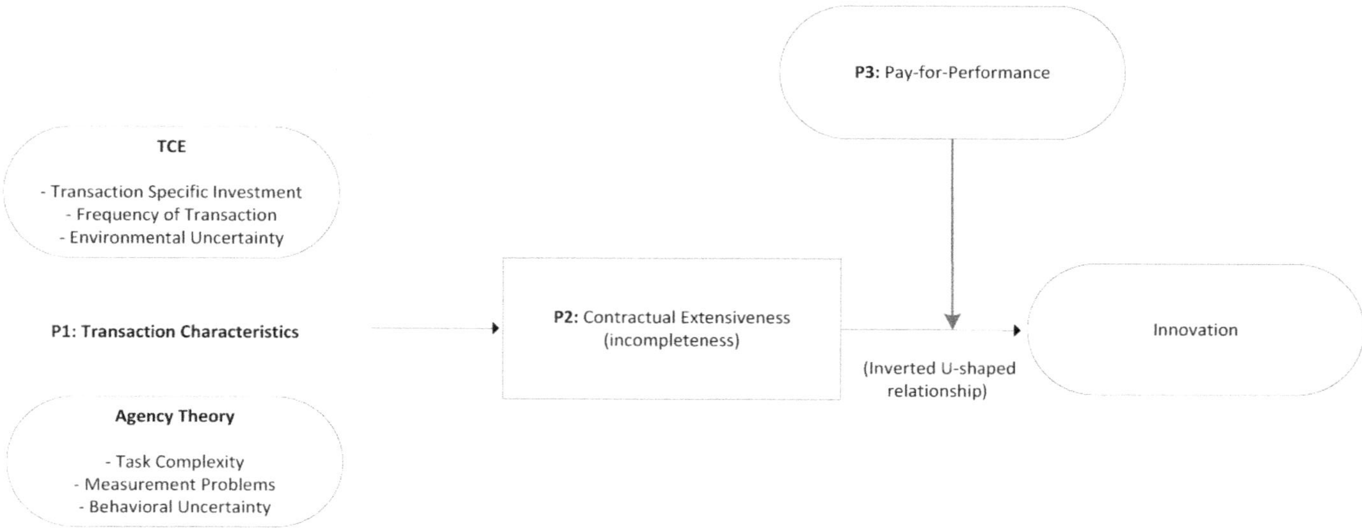

Figure 5.1. Conceptual model.

environmental uncertainty, measurement problems, task complexity, and behavioral uncertainty, affect the contractual degree of contractual (in)completeness. We argue that contractual incompleteness enables the partner to take its own decisions regarding how to deliver the transaction and which resources to use. Contractual incompleteness grants autonomy, which allows the partner freedom to engage in innovative activities. We do, however also acknowledge that a very high degree of incompleteness may lead to negative outcomes. This potential negative effect however can be mitigated by linking rewards more clearly to performance. Following AT-logic, linking rewards to performance incentivizes the partners to behave in the interest of the overall collaboration and allows them to use innovative actions in order to gain higher levels of performance. In sum, we argue that the negative effects of contractual incompleteness on innovative behavior can be mitigated by linking rewards to performance, in particular in the case of contractual incompleteness.

Our study extends research based on transaction cost economics and agency theory in the following way. In contrast to prior literature we have demonstrated the usefulness of combining TCE and AT in the context of contracts and innovation. We show that decisions in an IOR regarding contractual structure should not only consider exchange characteristics (as per TCE and AT), but also consider the effects of the degree of contractual (in)completeness resulting from these exchange characteristics on positive relationship outcomes, such as innovation. As such, our study addresses the scarcity of research on the effects of formal governance on performance and relational outcomes (Malhotra & Lumineau, 2011). Another implication of our research concerns the (positive and negative) effects of contractual incompleteness on innovation: the relationship between contractual incompleteness and innovation follows an inverse U-shaped relationship. This relationship posits that too little contractual incompleteness (i.e., too much contractual extensiveness) causes rigidity and curbs innovation, and too much contractual incompleteness provides the partner the possibility to act opportunistically. Though this suggests that there is a clear inflection point after which the effect of incompleteness on innovation becomes negative, we argue that potential negative effects can be mitigated by providing clear financial rewards (as suggested by AT). Thus, while TCE argues that the potential for opportunistic behavior as a result from a high degree of contractual incompleteness calls for a more complete contract, AT postulates to leave the level of contractual incompleteness as is, and to resolve potential opportunistic threats by pay- for-performance incentives. As such, TCE and AT suggest different solutions for the same problem. This observation provides support for the need to consider the two theories collectively rather than in isolation.

In addition, our results are also relevant to practitioners. Our findings firstly imply that contracts should leave room for the partners to discover new ways of working that eventually prove to be more effective and efficient. This can be achieved by leaving actions and solutions open (i.e., contractual incompleteness). This means that managers, when fostering innovation in IORs, need to move away from the detailed description of the transaction to guiding partner behavior. This way of managing IORs demands different capabilities from managers, who usually favor drafting contracts with high levels of detail to stay in control. However, obviously managers should realize that too much autonomy might lead to opportunistic behavior. In these cases, managers should incorporate financial incentives that reward performance outcomes in the design of their contracts. As such, linking rewards to performance provides a second solution for managers to move partner relationships towards innovation.

While our study shows interesting results, it also has limitations. First, we have concentrated on singling out the effects of formal control on innovation. We have not paid attention to other important factors that affect the relationship between contracting and innovation. Future research might also take the effects of relational governance attributes such as trust, communication and commitment into account (Mohr & Spekman, 1994; Morgan & Hunt, 1994). These specific attributes are also likely to impact the contracting-innovation relationship. A second limitation is that we have focused on innovation as the main relationship outcome. One specific contract design may however affect different performance measures differently. Therefore we would like to see more research using multiple performance indicators. Comparing the effects of contracts on innovation versus other outcomes would be an interesting future avenue.

Our research provides some other interesting future research opportunities as well. First, since we postulate that there is an optimal level of contractual incompleteness in relation to innovation, and that financial incentive schemes can attenuate the negative relationship between a high degree of contractual incompleteness and innovation, an interesting future avenue would be to determine the optimal level of contractual incompleteness. Another promising avenue is to research how specific pay-for-performance incentives affect innovation. We have only generally argued that pay-for-performance mechanisms will mitigate the negative effects of contractual incompleteness on innovation when there is a high degree of contractual incompleteness. It is likely that different pay-for-performance schemes show different results in terms of how they affect the relationship between contracting and innovation. Opening up the "black box" of performance incentives might be a next step towards the maturing of this model. Finally, our findings should be considered in relation to possible contingencies like, for example, service type and type of industry, and the nature of the transaction.

This research provides a valuable first step towards managing interorganizational relationships through an incomplete contract by postulating how the degree of contractual (in)completeness affects (positive) relationship outcomes.

REFERENCES

Abbey, A., & Dickson, J. W. (1983). R&D work climate and innovation in semiconductors. *Academy of Management Journal, 26*, 362-368.

Aghion, P., & Holden, R. (2011). Incomplete contracts and the theory of the firm: What have we learned over the past 25 years? *Journal of Economic Perspectives, 25*, 181-197.

Al-Najjar, N. I. (1995). Incomplete contracts and the governance of complex contractual relationships. *American Economic Review, 85*, 432-436.

Anderson, E. (1985). The salesperson as outside agent or employee: A transaction cost analysis. *Marketing Science, 4*, 234-254.

Anderson, S., & Dekker, H. C. (2005). Management control for market transactions: The relation between transaction characteristics, incomplete contract design, and subsequent performance. *Management Science, 51*, 1734-1752.

Arad, S., Hanson, M. A., & Schneider, R. J. (1997). A framework for the study of relationships between organizational characteristics and organizational innovation. *Journal of Creative Behavior, 31*, 42-58.

Argyres, N., Bercovitz, J., & Mayer, K. (2007). Complementarity and evolution of contractual provisions: An empirical study of IT services contracts. *Organization Science, 18*, 3-19.

Argyres, N., & Mayer, K. J. (2007). Contract design as a firm capability: An integration of learning and transaction cost perspectives. *Academy of Management Review, 32*, 1060-1077.

Barthelemy, J., & Quelin, B. (2006). Complexity of outsourcing contracts and ex post transaction costs: An empirical investigation. *Journal of Management Studies, 43*, 1775-1797.

Bernheim, D. B., & Whinston, M. D. (1998). Incomplete contracts and strategic ambiguity. *American Economic Review, 88*, 902-932.

Brousseau, E., & Coeurderoy, R. (2005). The governance of intellectual property rights in knowledge transfers: An empirical analysis of supervision provisions in technology licensing agreements. *International Journal of Economics and Business, 12*, 403-424.

Chen, Y., & Bharadwaj, A. (2009). An empirical analysis of contract structures in IT outsourcing. *Information Systems Research, 20*, 484-506.

Chesbrough, H. (2003). *Open innovation: The new imperative for creating and profiting from technology.* Boston, MA: Harvard Business School Press.

Chesbrough, H., Vanhaverbeke, W., & West, J. (2006). *Open innovation: Researching a new paradigm.* Oxford, UK: Oxford University Press.

Christensen, C. M., & Overdorf, M. (2000). Meeting the challenge of disruptive change. *Harvard Business Review, 78*(2), 66-76.

Coase, R. H. (1937). *The nature of the firm. Economica, 4(*16), 386-405.

Crocker, K. J., & Masten, S. E. (1991). Pretia ex machina? Prices and process in long-term contracts. *Journal of Law & Economics, 34*(1), 69-99.

Das, T. K., & Teng, B. (2001). Trust, control, and risk in strategic alliances: An integrated framework. *Organization Studies, 22*, 251-283.

Deckop, J. R., Mangel, R., & Cirka, C. C. (1999). Getting more than you pay for: Organizational citizenship behavior and pay-for-performance plans. *Academy of Management Journal, 42*, 420-428.

Deeds, D. L., & Rothaermel, F. T. (2003). Honeymoons and liabilities: The relationship between age and performance in research and development alliances. *Journal of Product Innovation Management, 20*, 468-485.

Devers, C. E., Cannella, A. A., Reilly, G. P., & Yoder, M. E. (2007). Executive compensation: A multidisciplinary review of recent development. *Journal of Management, 33*, 1019-1072.

Dooley, L., & O'Sullivan, D. (2007). Managing within distributed innovation networks. *International Journal of Innovation Management, 11*, 397-416.

Dyer, J. H., & Singh, H. (1998). The relational view: Cooperative strategy and sources of interorganizational competitive advantage. *Academy of Management Review, 23*, 660-679.

Eisenhardt, K. (1985). Control: Organizational and economic approaches. *Management Science, 31*, 134-149.

Eisenhardt, K. M. (1989). Agency theory: An assessment and review. *Academy of Management Review, 14*, 57-74.

Faems, D., Janssens, M., Madhok, A., & Van Looy, B. (2008). Toward an integrative perspective on alliance governance: Connecting contract design, trust dynamics, and contract application. *Academy of Management Journal, 51*, 1053-1076.

Faems, D., Van Looy, B., & Debackere, K. (2005). Interorganizational collabotion and innovation: Toward a portfolio approach. *Journal of Product Innovation, 22*, 238-250.

Ghoshal, S., & Moran, P. (1996). Bad for practice: A critique of the transaction cost theory. *Academy of Management Review, 21*, 13-47.

Goes, J. B., & Park, S. H. (1997). Interorganizational links and innovation: The case of hospital services. *Academy of Management Journal, 40*, 673-696.

Goldberg, V. P. (1976). Toward an expanded economic theory of contract. *Journal of Economic Issues, 10*(1), 45-61.

Goldberg, V. P. (1985). Price adjustment in long-term contracts. *Wisconsin Law Review*, May-June, 527-543.

Gulati, R., Lawrence, P. R., & Puranam, P. (2005). Adaptation in vertical relationships: Beyond incentive conflict. *Strategic Management Journal, 26*, 415-440.

Guth, W. D., & MacMillam, I. C. (1986). Strategy implementation versus middle management self-interest. *Strategic Management Journal, 7*, 313-327.

Hagedoorn, J. (2002). Inter-firm R&D partnerships: An overview of major trends and patterns since 1960. *Research Policy, 31*, 477-492.

Hamel, G. (1991). Competition for competence and inter-partner learning within international strategic alliances. *Strategic Management Journal, 12*(Summer), 83-103.

Hardy, C., Phillips, N., & Lawrence, T. (2003). Resources, knowledge and influence: The organizational effects of interorganizational collaboration. *Journal of Management Studies, 40*, 321-347.

Hart, O., & Moore, J. (1988). Incomplete contracts and renegotiation. *Econometrica, 56*, 755-785.

Heide, J. B. (1994). Interorganizational governance in marketing channels. *Journal of Marketing, 58*(1), 71-85.

Huston, L., & Sakkab, N. (2006). Connect and develop: Inside Proctor & Gamble's new model for innovation. *Harvard Business Review, 84*(3), 58-66.

Im, G., & Rai, A. (2008). Knowledge sharing ambidexterity in long-term interorganizational relationships. *Management Science, 54*, 1281-1296.

Jensen, M., & Meckling, W. (1976). Theory of the firm: Managerial behavior, agency costs, and ownership structure. *Journal of Financial Economics, 3*, 305-360.

Johnson, W. H. A., & Medcof, J. W. (2007). Motivating proactive subsidiary innovation: Agent-based theory and socialization models in global R&D. *Journal of International Management, 13*, 472-487.

Kim, S., Cohen, M. A., & Netessine, S. (2007). Performance contracting in after-sales service supply chains. *Management Science, 53*, 1843-1858.

King, D. R., Covin, J. G., & Hegarty, H. W. (2003). Complementary resources and the exploitation of technological innovations. *Journal of Management, 29*, 589-607.

Kloyer, M., & Scholderer, J. (2012). Effective incomplete contracts and milestones in market-distant R&D collaboration. *Research Policy, 41*, 346-357.

Leiblein, M. J., & Miller, D. J. (2003). An empirical investigation of transaction and firm level influences on the vertical boundaries of the firm. *Strategic Management Journal, 24*, 839-859.

Leonard-Barton, D. (1995). *Wellsprings of knowledge: Building and sustaining the sources of innovation*. Boston, MA: Harvard Business School Press.

Lumineau, F., & Malhotra, D. (2011). Shadow of the contract: How contract structure shapes interfirm dispute resolution. *Strategic Management Journal, 32*, 532-555.

Luo, Y. (2002). Contract, cooperation, and performance in international joint ventures. *Strategic Management Journal, 23*(, 903-919.

Lyons, B. L. (1996). Empirical relevance of efficient contract theory: Inter-firm contracts. *Oxford Review of Economic Policy, 12*(4), 27-52.

Makri, M., Lane, P. J., & Gomez-Mejia, L. R. (2006). CEO incentives, innovation, and performance in technology-intensive firms: A reconciliation of outcome and behavior-based incentive schemes. *Strategic Management Journal, 27*, 1057-1080.

Malhotra, D., & Lumineau, F. (2011). Trust and collaboration in the aftermath of conflict: The effects of contract structure. *Academy of Management Journal, 54*, 981-998.

Masten, S. E. (1993). Transaction costs, mistakes, and performance: Assessing the importance of governance. *Managerial and Decision Economics, 14*(2), 119-129.

Mayer, K. J., & Argyres, N. (2004). Learning to contract: Evidence from the personal computer industry. *Organization Science, 15*, 394-410.

Mintzberg, H. (1994). *The rise and fall of strategic planning*. Hemel Hempstead, UK: Prentice-Hall.

Mohr, J., & Spekman, R.(1994). Characteristics of partnership success: Partnership attributes, communication behavior, and conflict resolution techniques. *Strategic Management Journal, 15*, 135-152.

Mooi, E. A., & Ghosh, M. (2010). Contract specificity and its performance implications. *Journal of Marketing, 74*(2), 105-120.

Morgan, R. M. & Hunt, S. D. (1994). The commitment-trust theory of relationship marketing. *Journal of Marketing, 58*(3), 20-38.

Morris, M., Kuratko, D., & Covin, J. G. (2007). *Corporate entrepreneurship & innovation*. Mason, OH: South-Western CENGAGE Learning.

Ng, I. C. L., & Nudurupati, S. S. (2010). Outcome-based service contracts in the defence industry—mitigating the challenges. *Journal of Service Management, 21*, 656-674.

Ng, I. C. L., Maull, R., & Yip, N. (2009). Outcome-based contracts as a driver for systems thinking and service-dominant logic in service science: An evidence from the defense industry. *European Management Journal, 27*, 377-387.

Paolillo, J. G., & Brown, W. B. (1978). How organizational factors affect R&D innovation. *Research Management, 21*(7), 12-15.

Phelps, C. C. (2010). A longitudinal study of the influence of alliance network structure and composition on firm exploratory innovation. *Academy of Management Journal, 53*, 890-913.

Poppo, L, & Zenger, T. (2002). Do formal contracts and relational governance mechanisms function as substitutes or complements? *Strategic Management Journal, 23*, 707-725.

Prahalad, C., & Hamel, G. (1990). The core competence of the corporation. *Harvard Business Review, 68*(3), 79-91.

Reuer, J., & Arino, A. (2007). Contractual renegotiations in strategic alliances. *Journal of Management, 28*, 17-68.

Ring, P. S., & Van de Ven, A. H. (1992). Structuring cooperative relationships between organizations. *Strategic Management Journal, 13*, 483-498.

Roy, S., Sivakumar K., & Wilkinson I. F. (2004). Innovation generation in supply chain relationships: A conceptual model and research propositions. *Journal of the Academy of Marketing Science, 32*(1), 61-79.

Sampson, R. C. (2004). The cost of misaligned governance in R&D alliance. *Journal of Law, Economics, & Organization, 20*, 484-526.

Saussier, S. (2000). Transaction costs and contractual incompleteness: The case of Électricité de France. *Journal of Economic Behavior and Organization, 42*, 189-206.

Shepherd, D. A., & DeTienne, D. R. (2005). Prior knowledge, financial reward, and opportunity identification. *Entrepreneurship Theory and Practice, 29*(1), 91-112.

Shimizu, K. (2012). Risks of corporate entrepreneurship: Autonomy and agency issues. *Organization Science, 23*, 194-206.

Stump, R. L., & Heide, J. B. (1996). Controlling supplier opportunism in industrial relationships. *Journal of Marketing Research, 33*, 431-441.

Teece, D. J., Pisano, G., & Shuen, A. (1997). Dynamic capabilities and strategic management. *Strategic Management Journal, 18*, 509-533.

Tirole, J. (1999). Incomplete contracts: Where do we stand? *Econometrica, 67*, 741-781.

Tsai, W. (2001). Knowledge transfer in interorganizational networks: Effects of network position and absorptive capacity on business unit innovation and performance. *Academy of Management Journal, 44*, 996-1004.

van Echtelt, F. E. A., Wynstra, F., van Weele, A. J., & Duysters, G. (2008). Managing supplier involvement in new product development: A multiple-case study. *Journal of Product Innovation Management, 25*, 180-201.

Vanhaverbeke, W., Van de Vrande, V., & Chesbrough, H. (2008). Understanding the advantages of open innovation practices in corporate venturing in terms of real options. *Creativity and Innovation Management, 17*(4), 251-258.

Verbeke, A., & Greidanus, N. (2009). The end of the opportunism versus trust debate: Bounded reliability as a new envelope-concept in research on MNE governance. *Journal of International Business Studies, 40*, 1471-1495.

Wang, L., Yeung, J., & Zhang, M. (2011). The impact of trust and contract on innovation performance: The moderating role of environmental uncertainty. *International Journal of Production Economics, 134*(1), 114-122.

Williamson, O. (1979). Transaction-cost economics: The governance of contractual relations. *Journal of Law & Economics, 22*(2), 628-652.

Williamson, O. (1985). *The economic institutions of capitalism.* New York, NY: Free Press.

Zhou, Z. K., & Poppo, L. (2010). Exchange hazards, relational reliability, and contracts in China: The contingent role of legal enforceability. *Journal of International Business Studies, 41*, 861-882.

CHAPTER 6

ALLIANCE PORTFOLIO EVOLUTION IN THE CONTEXT OF SERVICE INNOVATION

The Case of ABN AMRO Dialogues Incubator

Marc D. Bahlmann, Ard-Pieter de Man, Brian V. Tjemkes, Alexander S. Alexiev, Jaspar Roos, and Paul L. Iske

ABSTRACT

This chapter reports on an alliance portfolio evolution case study in the context of service innovation. Extant literature mainly focuses on alliance portfolios at one point in time and has proposed two analytical approach es: path dependence and lifecycle. We propose that these two approaches should be used jointly while incorporating a third, namely, a co-evolutionary approach. We show the added value of this more comprehensive approach by examining ABN AMRO Dialogues Incubator whose alliance portfolio has gone through various iterations depending on the type of innovations it was meant to generate. We conclude with an alliance portfolio evolution model

Strategic Alliances for Innovation and R&D, pp. 133–157

consisting of the exogenous and endogenous mechanisms of path dependence, lifecycle stages, and co-evolution.

INTRODUCTION

In today's competitive business landscape, firms from both service- and product-based industries rely heavily on strategic alliances for an array of innovation objectives (Gulati, 1998; Tjemkes, Vos, & Burgers, 2012). Managing many strategic alliances simultaneously allows the focal firm to enjoy access to a variety of resources and markets. At the same time, it puts a heavy burden on the focal firm as interdependencies need to be managed purposefully and proactively (Anand & Khanna, 2000; George, Zahra, Wheatley, & Khan, 2001; Hoffmann, 2007; Lavie, 2007). Not surprisingly, the alliance portfolio phenomenon, here defined as all alliances of the focal firm, receives substantial scholarly attention.

Adopting the portfolio as a level of analysis represents a fundamental shift in alliance research (Gulati, 1998). The alliance portfolio perspective uncovers the possibility to touch upon a range of strategic issues not addressed by research on the antecedents and consequences of single, dyadic alliances. Essentially, alliance portfolios reflect a firm's current and past strategic orientation, which suggests that they are powerful and necessary strategic tools (Hoffmann, 2007; Lavie & Singh, 2011). Consequently, firms find themselves continually altering their alliance portfolio as their strategic and innovative orientation develops (Dittrich, Duysters, & De Man, 2007).

Extant literature has particularly focused on alliance portfolio emergence, management, and configuration (Wassmer, 2010). Much insight has been developed in these three domains respectively. However, as Wassmer (2010) notes, this insight is mainly the result of cross-sectional approaches that incorporate only the firm level of analysis, implying a bias toward endogenous, firm specific factors mainly. As a result, the bulk of studies view the alliance portfolio as a static phenomenon, and, in so doing, assume that portfolios are the result of rational, strategic processes (Wassmer, 2010), whereas literature exists that suggests otherwise (Lavie & Singh, 2011; Reuer & Ragozzino, 2006). Research is needed that focuses on alliance portfolio *evolution*, with an eye for its antecedents (Ozcan & Eisenhardt, 2009). Second, exploring the domain of alliance portfolio development and change requires a longitudinal approach, incorporating the potential influence of external factors.

The current study increases the understanding of alliance portfolio evolution by focusing on its drivers. Findings from a longitudinal, single case study performed at ABN AMRO Dialogues Incubator (henceforth Dialogues), a standalone venturing division of ABN AMRO Bank, suggest that alliance portfolio evolution is influenced by both a firm's endogenous and exogenous environment, and is driven by (1) path-dependent, (2) lifecycle-based, and (3) co-evolutionary dynamics.

This chapter is structured as follows: the theoretical framework provides an overview of current alliance portfolio research. This is followed by a contextualization of our case study, and continues with a discussion of our main findings. Finally, we discuss limitations and future research opportunities.

THEORETICAL FRAMEWORK

Since Gulati's seminal 1998 article on adopting a holistic approach when considering the role of alliances in firm performance, research on alliance portfolios has proliferated (Wassmer, 2010). The value of an alliance portfolio lies in the notion that it allows the focal firm to focus on its core competences, while benefiting from partners' resources and market positions (Yoffie & Slind, 2006) on the one hand, and dealing with strategic uncertainty on the other (Eisenhardt & Schoonhoven, 1996; Ozcan & Eisenhardt, 2009). Apart from gaining access to a wide variety of resources and knowledge (Powell, Koput, & Smith-Doerr, 1996), alliances also allow for risk mitigation and cost sharing. In all, a firm's alliance portfolio can be considered a crucial tool in its quest for reaching strategic goals (Sarkar, Aulakh, & Madhok, 2009; Schilke & Goerzen, 2010) and related performance outcomes.

Prior research has suggested that firms may proactively design their alliance portfolios in accordance with specific strategic goals. For instance, it has been established that it pays to approach prominent partners simultaneously when seeking high-performing portfolios (Ozcan & Eisenhardt, 2009). Also, it makes sense to balance cohesive ties with bridging ties, especially in view of innovation performance (Padula, 2008), and to allow for certain levels of functional and industry diversity (Jiang, Tao, & Santoro, 2010). In terms of portfolio management, prior studies indicate that firms differ in the extent to which they exhibit alliance coordination capabilities (Anand & Khanna, 2000; Khanna, 1998), and show the importance of understanding alliance portfolio interdependencies, for instance in terms of competition across portfolio partners (Lavie, 2007). Investing in alliance portfolio capabilities allows management to develop know-how and routines, for instance through dedicated management training (Bamford, Gomes-Casseres, & Robinson, 2003; Heimeriks, Klijn,

& Reuer, 2009; Lorenzoni & Lipparini, 1999; Schilke & Goerzen, 2010; Simonin, 1997).

Little attention has been paid to how and why firms change their alliance portfolio configuration over time (Lavie & Singh, 2011; Wassmer, 2010, p. 162). This is not to say that the topic of alliance portfolio evolution hasn't received any attention so far, but as Lavie and Singh (2011) noticed, it has mainly been driven by, "endogenous path-dependent alliance formation decisions" (p. 765) that take a unidirectional causal effect on portfolio evolutionary trajectories, where decisions to alter the alliance portfolio is guided by a combination of internal inducements and the availability of alliance opportunities. The current study adopts a wider perspective by also taking into account drivers of change external to the focal firm while moving away from unidirectional causal explanations of portfolio evolution.

The above implies that the current study takes a holistic approach to alliance portfolio change (Koza & Lewin, 1998), which allows for describing how alliance portfolio evolution and external stimuli (e.g., instances occurring in the firm's external environment) co-evolve (Hoffmann, 2007), while recognizing the potential influence of path dependent and lifecycle related mechanisms that drive portfolio evolution. Accordingly, developmental processes taking place at the level of the firm, alliance portfolio, and external environment may influence one another over time (Hoffmann. 2007). Following Lavie and Singh (2011), the present study seeks to advance a "contextualized account of specific causal linkages between a firm and its alliance portfolio" (p. 766), incorporating a firm's strategic decisions, innovative orientation, and its competitive, institutional and organizational environments.

Endogenous and Exogenous Drivers of Alliance Portfolio Evolution

Extant literature already recognizes drivers of change that can be considered both endogenous and exogenous to the focal firm and its alliance portfolio. Two main explanations stand out: (1) structural path dependence and (2) lifecycle stages (Lavie & Singh, 2011). These approaches are complemented by a third, namely co-evolution (Lavie & Singh, 2011). All three can manifest in the focal firm's endogenous and exogenous environments. With the first, Gulati (1995) suggested that a firm's decision to engage in new alliances is based on the accumulation of its prior alliance ties (both direct and indirect), forming a social context in which it is embedded. As such, firm amalgamation can be explained by factors stemming from social network and strategic interdependency perspectives, indicating the importance of social context as a catalyst of economic action (Granovetter, 1985). From a longitudinal perspective, Gulati (1995) finds that "the dynamics between social structure and alliance formation suggest a dialectic between action and social structure such that each influences the other" (p. 646). Firms rely heavily on information from their network of prior alliances when determining with whom to ally (Gulati & Gargiulo, 1999).

The second approach takes a lifecycle perspective, focusing on how alliance portfolios develop over time, focusing on configurative evolution as a result of firm action and coordination. For instance, Dyer and Nobeoka (2000, p. 360) suggested that Toyota created and managed a knowledge-sharing network through three so-called institutional innovations, namely (1) creating a supplier association, (2) making knowledge transfer consultants available free of charge, and (3) instigating small-group learning teams across the knowledge network. In so doing, Toyota produced a network that evolved "from weak ties, to strong bilateral ties with the convener (Toyota), to strong multilateral ties among suppliers" (p. 361). Likewise, Hite and Hesterly (2001) suggested that a firm's network will evolve depending on changing resource needs and challenges, developing from being primarily dense, cohesive, and path-dependent at a firm's emergence, into one that consists of both embedded and arm's-length ties, structural holes, and that is managed intentionally. In addition, Rothaermel and Deeds (2004) studied how technology ventures' alliance portfolios develop from an explorative to an exploitative orientation following a product development logic.

Third, Lavie and Singh (2011), when studying the case of Unisys Corporation, a U.S.-based technology company, found four interacting processes that influenced Unisys' alliance portfolio, namely (1) *co-evolution*, referring to the process of adjusting the portfolio as a result of changes in the focal firm's strategic orientation, (2) *external stimuli*, referring to drivers of the (co-)evolutionary process referred to above, such as changes in the technological environment, (3) *inertial pressures*, referring to internal crises and resistance to change, and 4) *restructuring and realignment*, referring to organizational changes that would realign Unisys' alliance portfolio configuration and strategy (Lavie & Singh, 2011).

As such, extant literature seems to suggest that alliance portfolio evolution drivers stem from the focal firm's internal and external environment, materializing through path-dependency, lifecycle stages, and co-evolution, thereby showing an intricate interaction among external, organizational, and strategic stimuli, with a pivotal role for those stemming from the external environment. Whereas the Unisys case refers mainly to the influence of the technological environment, prior literature has shown a great diversity of potential environmental influences and related strategic uncertainties.

Strategic uncertainty may stem from the regulative, technological, market or competitive environment (Hoffmann, 2007); it will be noticed that perceived uncertainty is what drives decision behavior (Huber & Daft, 1987). In relation to this, Hoffmann (2007) and Folta (1998) made a distinction between exogenous and endogenous uncertainty. Hoffmann found that the combination of perceived endogenous and exogenous uncertainty can lead companies to either engage in new explorative alliances, especially in the case of high strategic uncertainty, or exploitative alliances in the case of low strategic uncertainty. This synthesizes both endogenous and exogenous drivers of alliance portfolio evolution (see Table 6.1).

Table 6.1. Integrating Endogenous and Exogenous Perspectives of Alliance Portfolio Evolution

	Structural Path Dependence	*Lifecycle Approach*	*Co-evolution*
Endogenous perspective	Firms decide with whom to partner based on a combination of experience with former partners and information stemming from their current portfolio of partners.	Firms purposefully change their portfolio based on alterations in strategic objectives.	Alterations in the internal organization (e.g., structure, strategy) lead to portfolio evolution. Portfolio evolution may, in turn, instigate organizational change.
Exogenous perspective	External stimuli, such as changes in the technological and market environment, alter the social network in which firms are embedded and affect structural path dependence.	Strategic change follows external stimuli. The extent to which firms experience strategic uncertainty and shaping potential determines alliance portfolio evolution.	Environmental dynamics may affect alliance decisions at the micro-level. In turn, such micro-level decisions may influence the broader alliance environment.
Associated references	Gulati, 1995; Gulati & Gargiulo, 1999; Granovetter, 1995; Lavie & Singh, 2011.	Dyer & Nobeoka, 2000; Hite & Hersterly, 2001; Hoffmann, 2007.	Tjemkes et al., 2012; Volberda & Lewin, 2003.

The foregoing makes clear that alliance portfolio evolution must be considered from a range of perspectives. The remainder of this study will therefore focus on identifying endogenous and exogenous drivers, with particular emphasis on their interaction.

METHODS AND CASE DESCRIPTION

Since not much theory is available on the antecedents that drive alliance portfolio configuration, we conducted an inductive, single case study (Eisenhardt, 1989). An inductive approach is especially suitable for developing theoretical insights in an area that hasn't been addressed extensively before. Also, a case study approach is suitable when dealing with a process-like question (Yin, 2003).

Our research is positioned in the financial industry and specifically took place at ABN AMRO Dialogues Incubator. At the time of our study, Dialogues could be described as a standalone entity focused on generating and developing innovative financial industry projects and ventures. While so doing, Dialogues remained closely connected to ABN AMRO. Our choice for this specific case was appropriate for a number of reasons. First, Dialogues was established with the explicit purpose of building an alliance portfolio aimed at generating innovative projects of which the successful ones, in due time, could be leveraged to the mother company. Second, Dialogues' task was to leverage current relationships from the mother company, meaning that many interdependencies would occur. Third, Dialogues is a relatively young entity, officially dating from 2006, which made it easier to reconstruct its portfolio development from birth to present. Fourth, the case allowed us to investigate the influence of exogenous factors, since the financial industry is currently undergoing some radical changes.

Being accommodated in the former ABN AMRO trading floor, Dialogues' initial activities were aimed at "*generating, identifying, and accelerating*" innovations, as they would formulate it themselves. More specifically, Dialogues' employees conducted research, studied relevant trends and developments, engaged in new partnerships, all with the aim of creating innovative value for the mother company. In so doing, Dialogues applied what its CEO calls "combinatoric thinking" (Iske, 2010), an analogue to Schumpeterian (1934) innovation. In this, the innovation process itself passes through a number of distinct phases, called "stage-gating" (Iske, 2010), which is similar to the classic innovation funnel. All of this departs from the overarching vision of exploring the "*next generation bank,*" that is, an abstract, future representation of what a bank should look like, what its purpose should be in present and future society, and what future markets it should address. As such, it was Dialogues' objective to find "*sizeable new business*" and invoke change such that the mother company would be prepared for future financial industry developments.

Data Sources

In the process of reconstructing Dialogues' alliance portfolio, as well as its drivers, we made use of archives, interviews, informal follow-ups through e-mail and telephone, and desk research. By relying on multiple sources, and collecting the data periodically, we increased the accuracy of our findings and their internal validity (Gibbert, Ruigrok, & Wicki, 2008; Jick, 1979). The semistructured interviews proved an especially valuable data source. In total, we conducted seven interviews with actors throughout the organization (see Table 6.2 for an overview) in order to verify information across respondents, while gaining both strategic and operational insight into Dialogues' alliance portfolio. As such, we were able to gain a complete, accurate picture of the current case. To this end, we interviewed the company's CEO, managers, project managers, and an internet designer. Typically, interviews would range from 60 to 90 minutes, and were fully transcribed shortly afterwards.

Table 6.2. Respondent and Interview Characteristics

Respondent ID	*Job Description*	*Duration of Interview*
R1	CEO, Dialogues	45 minutes
R2	Chief Inspiration Officer, Dialogues	90 minutes
R3	Manager, Dialogues Venture	60 minutes
R4	Facility manager	60 minutes
R5	Project manager	60 minutes
R6	Director, Dialogues Holding	70 minutes
R7	Internet designer	85 minutes

The case study protocol consisted of three sections, namely a description of the research design and setting, an overview of relevant conceptual and operational definitions, and a semistructured interview guide. The guide was structured in six stages, as follows. First, the interview would start by explaining the background of the study and addressing potential confidentiality issues, allowing the respondent to talk about his or her expertise, professional experience, and company role. Stage two of the interview would address Dialogues' context and business environment, as well as its market orientation, role within the mother company, and general approach to innovation, important customers, and dominant product or service offerings. Stage three aimed at generating detailed information about Dialogues' alliance portfolio and was more focused on probing for potential additional partners, their selection rationale, the nature of the partnerships, and the importance of partners. Stage four addressed portfolio management, emphasizing approach, capability, interdependencies, contractual forms, performance measurement, and coordination efforts. Given Dialogues' main objective (that is, to innovate), stage five addressed innovation activities and related projects. This resulted in detailed information about their innovation approach, the numerous innovative ventures, the role of alliance partners, and associated costs and drawbacks. Stage six finalized the interview by asking for additional information (for example, minutes of meetings, PowerPoint presentations, and miscellaneous archival sources).

Data Analysis

Data analysis began by transcribing all interviews verbatim with a specific focus on discovering inadequacies or contrasting views. This allowed for triangulating respondent claims (Jick, 1979), as well as verifying respondent consistency. We dealt with inconsistent findings or ambiguities by follow-up e-mails and telephone conversations. Subsequently, a case report described findings for each of the interview topics described above, and presented the first analysis of key insights. This was then reviewed by a number of knowledgeable informants (that is, the CEO and CIO) on remaining inconsistencies and misinterpretations, for instance, with respect to the ordering of major events.

FINDINGS

Figure 6.1 offers a longitudinal overview of events that played a role in Dialogues' alliance portfolio evolution and will therefore serve as our guide in describing how a combination of exogenous and endogenous factors shaped Dialogues' alliance portfolio.

Dialogues' Takeoff: Between Exploration and Exploitation

From the very start in 2006, Dialogues was engaged in building an alliance portfolio in view of its core objective, namely to leverage ABN AMRO's partners (for example, Tata Consultancy Services (TCS), CapGemini, Achmea, IBM). Since its start, Dialogues has explored many —sometimes distant—markets through a range of innovative ventures and initiatives. The central thought was that Dialogues would function as ABN AMRO's venturing hub, providing the mother company with information

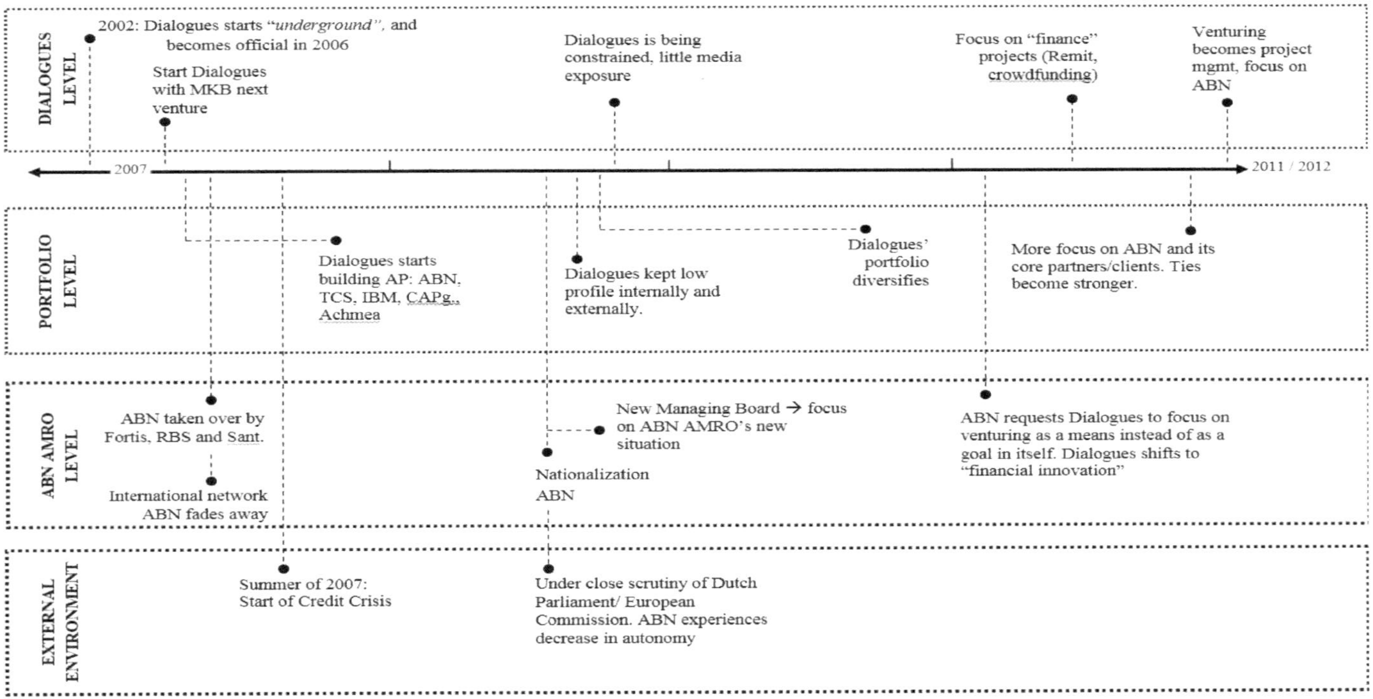

Figure 6.1. Timeline Dialogues.

on societal developments and trends. An additional objective was to start projects that would utilize existing mother company partnerships in a more innovative manner.

Dialogues' initial success came with the *MKBnext* venture, which aimed at facilitating business transfer in the domain of SMEs.[1] MKBnext was already in existence prior to Dialogues' official start, but was placed under its supervision, partly to emphasize Dialogues' launch, but also because it required specialized attention. In addition, MKBnext fit in nicely with Dialogues' prime mission of finding and developing sizeable new businesses. The MKBnext venture can be categorized as a service innovation that would adhere to March's (1991) exploitation dimension in the sense that it capitalized on expertise already available in the mother company (or its network), yet offered it to the market in a novel way. However, the bulk of innovative ventures following Dialogues' start would adhere to March's exploration dimension of innovation. Ventures such as Dialogues Technology, Brightbox, and Associates (see Table 6.3 for venture descriptions) involved a fair degree of knowledge exploration on the part of Dialogues, thereby engaging in "variety in experiences" (Holmqvist, 2003, p. 99) through experimentation and innovation.

The first phase of venturing can therefore be described as one where Dialogues' innovation orientation was skewed toward exploration (see Figure 6.2) rather than exploitation (March, 1991), in spite of Dialogues' initial dual innovative orientation. For instance, the Associates venture involved creating and exploiting a so-called online platform for freelancers to function as an intermediary with potential business clients. Likewise, Brightbox involved transferring human resources management best practices developed by the mother company to its major clients. In both cases, the ventures' service offerings did not reflect traditional banking services for which the mother company was traditionally known. Engaging with alliance partners that were cognitively distant skewed Dialogues' innovative orientation accordingly. Dialogues, and by extension the mother company, used such ventures to explore new markets and new knowledge domains. In both cases, Brightbox and Associates were ultimately sold, and now are alliance partners of ABN AMRO.

In this phase, Dialogues' portfolio could be described as primarily consisting of partners stemming from the mother company's network of clients and (preferred) suppliers (Table 6.4). Furthermore, it was relatively homogenous both in terms of partner organization type (relatively large organizations such as TCS, IBM, etc.) and industry background (mainly finance and IT). Dialogues' focus was mainly aimed at allying with key partners of the mother company. This resulted, for instance, in partnerships with TCS, IBM, CapGemini, and Sogeti that would ultimately indicate a move toward the explorative dimension (March, 1991).

Table 6.3. Description of Dialogues' Ventures 2007-2011

Venture	Description and objective
ArtStart	Online platform aimed at revolutionizing art trade.
Associates	Online platform aimed at facilitating freelance workers in their search for a client. By means of this functionality, Dialogues aimed to simplify the traditional value chain by taking in an intermediary position between freelancer and potential client.
Brightbox	HR management optimization tool. Its objective was to identify over- and under-capacity in terms of personnel at all mother company departments. Subsequently, the tool would allow company management to temporarily allocate manpower from departments characterized by overcapacity to those experiencing under-capacity.
Dialogues Technology	Venture specializing in interactive design, according to "Scrum" methodology.
Family Compass	Banking tool. Provides insight in spending and personal findings to banking customers.
FansFirst	Venture applied the logic behind option trading to the market for sporting events (cricket, soccer, et cetera).
iD Express	Venture aimed at connecting inventors to market parties.
MKB NEXT	Online platform aimed at facilitating business transfer of SMEs.
MyGoodWorks	Online platform, with the objective of connecting charity donors (that is, customers and clients) to charity organizations.
PPI	New venture developing new pension services in response to anticipated changes to Dutch pension/fiscal legislation.
Remittances	Alternative to MoneyTransfer.
Seeds	Venture aimed at exploring the possibility of crowd sourcing, taking into account Dutch legislation.

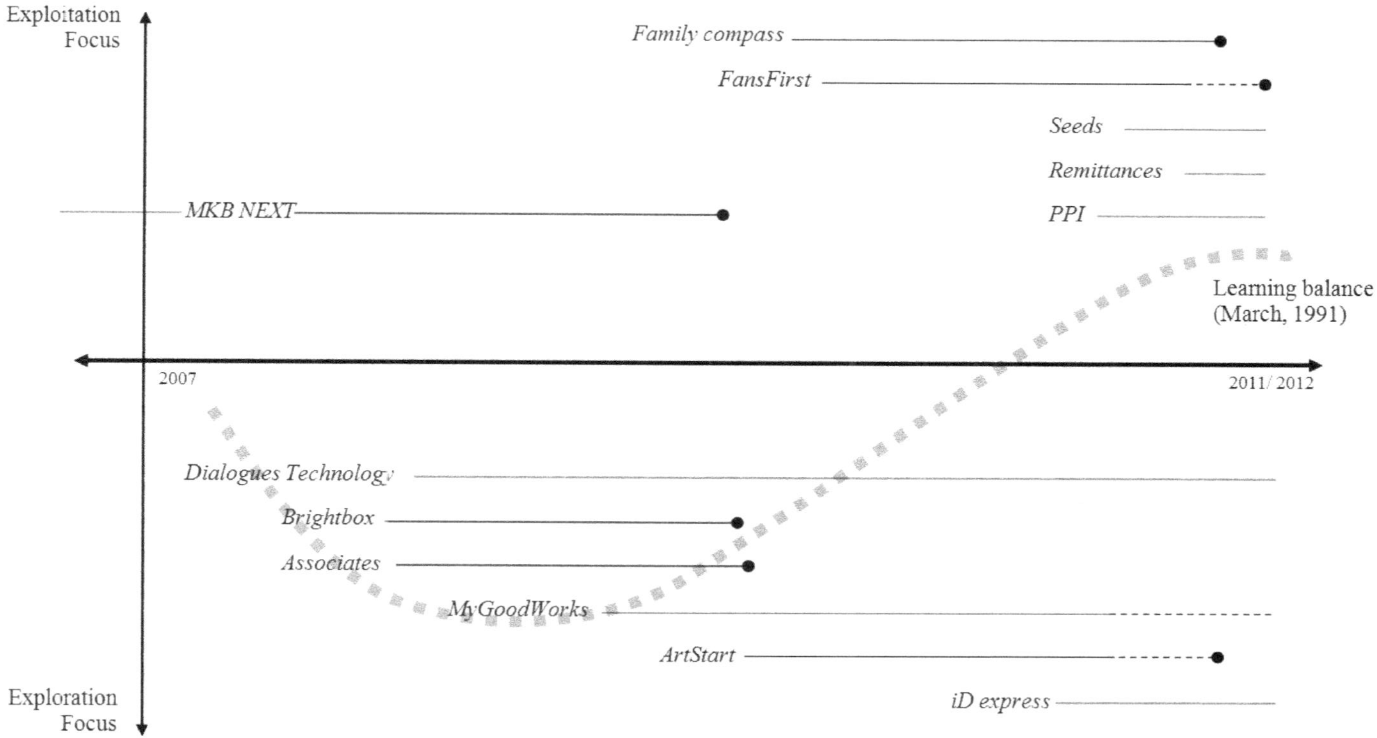

Figure 6.2. Overview of Dialogues' ventures 2007-2012.

Table 6.4. Alliance Portfolio, Dialogues, 2006-2012

	2006-2007	*2007-2010*	*2010-2012*
Portfolio Size	Small, start-up phase	Large, 55+ alliances	Medium, 20-25 Alliances
Portfolio diversity	Low	High	Moderate
	Mainly finance and IT-partners.	Many organizations from a range of industries are included in the portfolio, including organizations from the field of marketing, event organizing, charity, IT, creative industry, Internet.	More focus on exploitative innovation results in inclusion of ABN's corporate clients in the portfolio.
Portfolio partner type	Mother company mainly, in addition to some Multi-National Enterprises and preferred suppliers, ABN.	Further expansion of ties with MNEs and preferred suppliers in 2007. Over the course of 2008–2009, the portfolio diversifies significantly through incorporating many small and medium-sized partners.	Both medium- and large-sized companies. Stronger focus on partners of ABN AMRO. Incorporation of corporate clients.
Exemplary partners	ABN AMRO mother company, ABN AMRO HR, Philips, Achmea.	IBM, TCS, TEDX, universities, media companies, consulting firms, platform organizations, charity events and organizations, trend-watching firms, governmental organizations, legal firms, incubators.	IBM, Universities, corporate clients like Randstad, Large pension insurer, AAV insurance (among others).

Exogenous Interruptions: Hostile Takeover and Nationalization

Shortly after Dialogues' start, and its subsequent initiation of both explorative and exploitative ventures, the mother company faced challenging times. In the spring of 2007, a consortium consisting of three banks (Royal Bank of Scotland, Banco Santander, and Fortis Bank) announced that they would bid on ABN AMRO at around 39 euros a share, outbidding Barclays Bank, which was ABN AMRO's preferred takeover party. The consortium eventually won the bid, which split up ABN AMRO as an international banking corporation as it meant the annexation of ABN's American activities by the Royal Bank of Scotland (RBS), ABN's Brazilian and Italian[2] divisions by Banco Santander, and its Benelux activities by Fortis Bank.

The takeover struggle by RBS, Santander, and Fortis had vast implications for Dialogues' alliance portfolio activities. Suddenly, the mother company's international network of business relationships was inaccessible. As one respondent from Dialogues explains:

> ABN AMRO was, for instance, always very active in France, the United States, but now there are hardly any U.S.-based partners in our portfolio. Sure, IBM is in our portfolio, but that's because they are very active in the Netherlands. But if this hadn't been the case [that is, the takeover], than I would've been traveling to the U.S., Chicago, New York. But this never happened because the bank was sold, ... so we hardly have any American contacts anymore.

The takeover not only increased Dialogues' strategic uncertainty, but also limited its Benelux portfolio-building opportunities. This also meant a drastic cut in Dialogues' strategic and innovation opportunities. This was exacerbated once the 2007 credit crisis started to affect both Banco Santander and Fortis Bank. In the case of Fortis Bank, the credit crunch made painfully clear that it could not finance their part of the takeover (approximately €24 billion). After Fortis failed to find a new buyer for the ABN AMRO Benelux division, the Dutch government eventually decided to buy ABN AMRO (and Fortis' Dutch division) in order to prevent it from getting infected by the credit crisis (which took place in the fall of 2008, almost parallel to the Chapter 11 filing of Lehman Brothers).

The nationalization of ABN AMRO carried vast implications for Dialogues' strategy and positioning. Originally, the focus was on entrepreneurship, venturing, and outside-in innovation. However, the organization was in this period too busy solving urgent matters and meeting new requirements as a consequence of the new ownership structure, resulting in less time and attention for Dialogues' activities. In spite of this, Dialogues was allowed to move on, but with a somewhat lower profile and on a more self-propelling and self-steering basis. The new managing board, a mix of former board members and newcomers, had to focus on getting the organization back on track, but in the years that followed started to understand and appreciate Dialogues. As stated, in that period, Dialogues was autonomous to a large extent, with corresponding advantages (freedom to act) and disadvantages (limited access to ABN AMRO's network and capabilities).

Portfolio Evolution 2007-2010

This period can be described as one where Dialogues experienced high strategic uncertainty combined with low shaping potential (that is, the ability to actively shape or respond to environmental changes) (Hoffmann, 2007). It compelled Dialogues' management to build a network that would increase its shaping potential.

Following the exogenous interruptions described above, Dialogues' portfolio would grow and diversify immensely. This development continued throughout 2008 and 2009 with the incorporation of partners from a variety of industries, such as IT, media, marketing, the creative industry, and development aid. Whereas the initial steps in building Dialogues' portfolio involved leveraging the mother company's partners, the process of building a more diversified portfolio drew heavily on the personal and business networks of respondents R1 and R2.

> That is typical for [name R1] and [name R2]. These guys talk to everyone, have many external appointments. They bring in lots of new ideas and contacts.

A number of factors have contributed to this way of portfolio building. First, diversifying the portfolio took place in view of Dialogues' main objective of finding and developing new markets. Second, the exogenous stimuli that eventually would lead to Dialogues maintaining a low profile subsequently meant it was not allowed to contact senior executives from ABN's client network and preferred suppliers in the period following the nationalization. This further provoked Dialogues' management to start building an alliance portfolio characterized by many external, diverse contacts.

Through engaging with smaller, more distant partners, Dialogues' increasingly diverse portfolio could more thoroughly emphasize explorative service innovation. This can be illustrated by two ventures initiated during this phase: *MyGoodWorks* and *ArtStart*. MyGoodWorks involved creating an online platform that would allow novel ways for benefactors (that is, clients of ABN AMRO) to get involved in charity. ArtStart also involved an online platform, and aimed at intervening in the thus-far traditional way of buying art, namely through galleries and at exhibitions. Both ventures were explorative in the sense that they did not build on Dialogues' and ABN AMRO's traditional banking knowledge. This also meant that ABN's network did not suffice in realizing these ventures. Rather, new alliances needed to be established.

2010–2012: Strategic Change and Portfolio Evolution

The exogenous interruptions described above not only led ABN AMRO to focus more on its core banking activities, but also meant that the market for new ventures, one of the core activities of Dialogues, was challenging. The drastic reduction of mother company resources, combined with the worsened market conditions for new ventures and an increased desire to provide more tangible innovations, led Dialogues to focus its core activities more on creating value for ABN AMRO. As one respondent explains:

> So we maneuvered to the Bank more closely. Our alliance with [large Pension administrator] and the associated venture exemplifies this maneuver. Our venture in the art industry [that is, ArtStart], by contrast, wasn't close enough to ABN's core business. Crowdfunding, which we are involved in now, does relate to the Bank's core business.

Dialogues' strategic movement can be described as aimed at seeking more synergy between the mother and its subsidiary. In practice, this means that Dialogues' explorative innovation orientation became less prominent, while its exploitative innovation activities within the mother company increased. Ventures such as *Seeds* (crowd-funding), *Remittances* (money transfer services) and *PPI* (pension services) demonstrate this shift (see Figure 6.2). Following the hectic 2007-2010 period, Dialogues started to position itself as ABN AMRO's innovation "laboratory." Specifically, Dialogues' new task was to investigate exploitative innovation opportunities for the mother company and her clients. Thus, Dialogues' innovation expertise was being leveraged to create value for ABN's corporate clients. The focus shifted to "applied financial innovation", in order to stay connected to the mother company and prevent Dialogues from becoming irrelevant. In addition, innovation projects now take form as small proof of concepts rather than big international launches.

This reorientation resulted in a drastically changed alliance portfolio configuration in terms of size, partner type, and diversity (Table 6.4). First, the portfolio became smaller in the post-2010 phase. During the 2007-2010 phase, Dialogues' portfolio consisted of 55+ alliance partners, many of which were small organizations. Following the shift in innovation orientation, Dialogues' alliance portfolio decreased to about 20-25 partners. In terms of partner type, this also implied a shift toward incorporating ABN AMRO's corporate clients rather than building a standalone alliance portfolio. In addition, the 2010-2012 phase was characterized by moderate diversity, as the vast amount of differentiation among alliance partners was cut back; Table 6.4 summarizes this development.

DISCUSSION

The findings make clear that portfolio evolution stems from both exogenous and endogenous factors. In the case of Dialogues, we must conclude that exogenous factors stimulated innovation orientation and related partnering activities. Figure 6.3 illustrates the interacting factors leading to portfolio evolution.

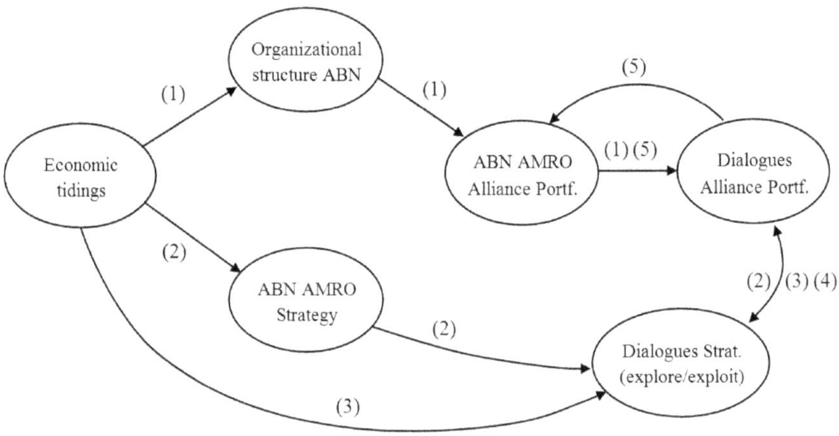

Figure 6.3. Model of alliance portfolio evolution based on the Dialogues case.

In the case of ABN AMRO, exogenous factors led to considerable changes in organizational structure and strategy. This not only affected the mother company in terms of structure and strategy, but directly affected Dialogues' alliance portfolio as well. The credit crunch that followed the takeover further complicated issues for both ABN AMRO and Dialogues, as ABN's internal strategy would be more exploitation oriented and the market for new ventures became increasingly more challenging. The Dialogues case distinguished three distinct exogenous processes that determine alliance portfolio evolution:

1. Exogenous stimuli affected ABN's organizational structure, which subsequently affected ABN AMRO's alliance portfolio, thereby limiting Dialogues in building its own alliance portfolio (indicated by path 1, Figure 6.3).
2. Exogenous stimuli affected ABN's strategy, which subsequently affected Dialogues' strategic orientation and associated alliance portfolio (indicated by path 2, Figure 6.3).

3. Exogenous stimuli affected Dialogues' strategy directly, which resulted in subsequent alterations to its portfolio of alliances (indicated by path 3, Figure 6.3).

Turning to the endogenous drivers of alliance portfolio evolution, the case study makes clear that Dialogues' portfolio structure was affected by changes in strategy, while, in turn, Dialogues' structure was affected by the configuration of its portfolio. Also, Dialogues' successes (e.g., Brightbox, Associates) would cause a change in the alliance portfolio of the mother company. Thus, based on the Dialogues case, we can distinguish two additional endogenous processes that determine alliance portfolio evolution:

4. Changes in Dialogues' alliance portfolio in turn affected Dialogues' strategy (indicated by path 4, Figure 6.3).
5. Successful innovation projects (Brightbox, Associates) would result in alterations and new alliances at ABN AMRO (indicated by path 5, Figure 6.3).

The above-described processes are integrated in Figure 6.3. Processes 1 and 2 underscore the limiting effect of interdependency between the mother company and Dialogues. First, Dialogues was affected by crucial changes in ABN AMRO's organizational structure; second, Dialogues was forced to alter its strategy in line with changes in the mother company's strategy. For Dialogues, this meant that it changed its predominately explorative orientation into an exploitative one. This is illustrated by the bulk of ventures that characterized Dialogues' venture portfolio following the main exogenous developments (that is, FansFirst, Seeds, PPI). From the perspective of Dialogues, this strategic change aimed at providing more value to the mother company. Process 3 stresses the effect of exogenous factors on alliance portfolio evolution, mediated by strategic change at the Dialogues level. Process 4 indicates the co-evolutionary mechanism, as diversifying the alliance portfolio in turn resulted in an increased focus on exploratory service innovation. Finally, process 5 underscores the co-evolution that takes place between different layers in the organization, or, in our case, between ABN AMRO on the one hand and Dialogues on the other.

The case study provides support for both the path-dependent and lifecycle approaches to explaining portfolio evolution, next to co-evolutionary dynamics. First, path-dependent drivers are noticeable as Dialogues relied heavily on already-established personal and business networks to develop its own alliance portfolio. Second, a lifecycle rationale is distinguishable as Dialogues' portfolio-building efforts progressed from seeking dense, strong relationships to a portfolio consisting of a bigger variety of ties

(Hite & Hesterly, 2001). Third, the Dialogues case also substantiates the co-evolutionary perspective. At its founding in 2006, Dialogues was given the task of producing innovations in both the explorative and exploitative domain (March, 1991), while turning preferred suppliers from the mother organization into alliances in order to leverage their capabilities. As Dialogues allied with ABN AMRO's preferred suppliers, an alliance portfolio emerged that was homogeneous in setup, yet consisted of partners with skills and knowledge very different from Dialogues. This particular configuration skewed its innovation orientation (that is, its strategy) toward exploration, thereby constraining Dialogues' intended ambidextrous innovative orientation. In addition, co-evolution took place between the corporate and business level, as ABN AMRO sought to enhance internal performance by allying with Brightbox and Associates. This latter notion suggests that portfolio co-evolution can take place both within and between analysis levels (as depicted in Figure 6.4).

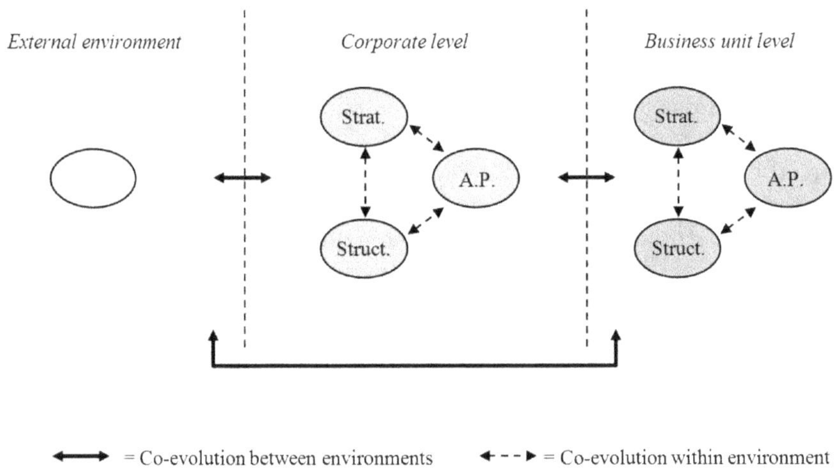

Figure 6.4. Conceptual model of co-evolution.

The case study findings offer ample opportunity for theoretical reflection. The Dialogues study exhibits significant alliance portfolio evolution in a relatively short period of time, where root drivers stem from the mother company's external environment as well as its own. In this case, portfolio evolution appeared to be reactive for the most part, following both exogenous forces and subsequent changes in strategy, manifesting through processes of path dependency, life cycle changes, and co-evolution.

In line with Lavie and Singh (2011), the present study underscores the exogenous factors that influence alliance portfolio change. In so doing, we complement the main two portfolio evolution explanations, that is, structural path dependence (Gulati, 1995; Gulati & Gargiulo, 1999) and lifecycle phases (e.g., Hite & Hesterly, 2001). Portfolio evolution patterns must be perceived holistically; that is, from both an endogenous and exogenous perspective, in order to understand their evolution trajectory. Crucial in this respect is the notion that the identified dynamics are interdependent (Lavie & Singh, 2011). To study alliance portfolio evolution from either an exogenous or endogenous perspective on the one hand, and only a path-dependent, lifecycle and co-evolutionary rationale on the other, is likely to result an incomplete representation of underlying dynamics.

MANAGERIAL IMPLICATIONS

The picture that emerges is of highly unpredictable dynamics affecting alliance portfolios. Portfolio management is therefore unlikely to be a planned and linear process. Instead, it will be emergent, unpredictable, and messy. This does not suggest that management's role is essentially reactive; rather, the Dialogues case shows three different tactics managers may deploy to ensure their portfolio continues to fit with corporate strategy.

The first tactic is to identify the need for change. Essentially this requires managers to bridge the internal organization to the external in the portfolio. For example, by building internal coalitions that support external partnering, the alliance portfolio will be less vulnerable to strategic changes. In addition, managers should monitor changing market and organization priorities so they can be proactive in changing their portfolio.

The second tactic is about improving portfolio change capability. The focus is less on creating a perfect portfolio at one point in time than on ensuring it can adapt to new demands. The Dialogues case shows how companies can achieve this. First, they can build less intensive or at least less formal alliances that are easier to disband or to change. When detailed long-term contracts form the basis of portfolio partnerships, they will be more difficult to restructure. Second, in addition to the formal alliances, Dialogues invested in an extensive informal network of familiar organizations. Each presented an option for a new alliance. An informal network ensures a flow of new partners and new ideas that may help to find new partners quickly when circumstances demand. Third, managing partner expectations about what they can expect from Dialogues deserves more attention. The core values signal that Dialogues is a dynamic business. The more partners understand the nature of Dialogues' business, the easier it will be to inculcate network change. Communicating those values to partners is an important management task.

A third tactic is to limit the risk that change will disrupt the entire network. The alliance portfolio consists of a mix of partners that partly come from ABN AMRO, but a portfolio of other partners has been built up as well. Coordination between partners across the portfolio is markedly different than in other alliance portfolios. For example, whereas Toyota strives to connect its partners as much as possible to foster learning and knowledge exchange (Dyer & Nobeoka, 2000), Dialogues mainly runs its portfolio as a set of unconnected bilateral relationships. This increases flexibility in two ways. First, by spreading risk, it is unlikely that a portfolio has to be changed completely. Second, portfolio reshuffling will be easier with bilateral partnerships than when all partners are intimately connected. In the latter case, change will slow down because it will have to involve multiple partners.

CONCLUSION, LIMITATIONS, AND FUTURE RESEARCH

This study set out to develop a better understanding of the drivers of alliance portfolio evolution. Findings of a case study performed at ABN AMRO Dialogues suggest that alliance portfolio evolution is the result of path-dependent lifecycle stages, and co-evolutionary processes that become manifest in endogenous and exogenous environments. Moreover, endogenous and exogenous drivers heavily interact in alliance portfolio evolution.

This study is limited due to the single case study setup. Although this approach allows for increased depth and theorization, it is at the expense of empirical robustness. In other words, the value of the current approach stems from its analytical generalization qualities; future work may attempt generalization instead (Gibbert et al., 2008), through replication of the above by quantitative means, or theoretical generalization by means of multiple case study designs. In addition, the current case study context is limited to a smaller entity focused on innovation while embedded in a multinational organization. As our inquiry was mainly directed at Dialogues, we might have overlooked corporate dynamics, as well as potential impacts stemming from the unit and directed to the mother company. The specificity of this case implies that any theoretical implications derived from it may not be generalizable to small and medium-sized companies, as well as business units that are not dedicated to innovation. Also, adopting a portfolio perspective assumes lower-level homogeneity at the level of dyadic alliances (Felin & Hesterly, 2007). The current focus on alliance portfolio configuration overlooks heterogeneity at the dyad level, meaning that some partners may affect outcomes significantly more than others. Regarding alliance portfolio management, the Dialogues case identified a number of tactics that differ markedly from those in stable business environments. Further research may shed light on the question of whether other tactics exist in dynamic environments and what the implications are for managerial agency. Large-scale research may next assess the effectiveness of various portfolio management tactics. Finally, the current case study has a relatively short timeframe of 6 years. When seeking to identify co-evolutionary processes, a larger timeframe is appropriate. Nonetheless, the present study demonstrates the intricate and intriguing nature of alliance portfolio evolution, exposing the need for a thorough re-evaluation of portfolio dynamics in general and evolution specifically.

NOTES

1. In Dutch, the abbreviation "MKB" stands for Midden- en Kleinbedrijf, which can be translated as Small and Medium-sized Enterprises, or "SME."
2. In Italy, ABN AMRO operated under the name of Antonveneta.

ACKNOWLEDGMENTS

This chapter is part of a larger research project on service innovation, which received funding from the Dutch "Pieken in de Delta" program. We thank James Morrison for text editing. Any mistakes in the manuscript are the sole responsibility of the authors.

REFERENCES

Anand, B. N., & Khanna, T. (2000). Do firms learn to create value? The case of alliances. *Strategic Management Journal, 21*, 295-315.

Bamford, J. D., Gomes-Casseres, B., & Robinson, M. (2003). *Mastering alliance strategy: A comprehensive guide to design, management, and organization*. San Francisco, CA: Jossey-Bass/Wiley.

Dittrich, K., Duysters, G., & De Man, A-P. (2007). Strategic repositioning by means of alliance networks: The case of IBM. *Research Policy, 36*, 1496-1511.

Dyer, J. H., & Nobeoka, K. (2000), Creating and managing a high-performance knowledge-sharing network: The Toyota case. *Strategic Management Journal, 21*, 345-367.

Eisenhardt, K. M. (1989). Building theories from case study research. *Academy of Management Review, 14,* 532-550.
Eisenhardt, K. M., & Schoonhoven, C. B. (1996). Resource-based view of strategic alliance formation: Strategic and social effects in entrepreneurial firms. *Organization Science, 7,* 136-150.
Felin, T., & Hesterly, W. S. (2007). The knowledge-based view, nested heterogeneity, and new value creation: Philosophical considerations on the locus of knowledge. *Academy of Management Review, 32,* 195-218.
Folta, T. B. (1998). Governance and uncertainty: The trade-off between administrative control and commitment. *Strategic Management Journal, 9,* 123-139.
George, G., Zahra, S. A., Wheatley, K. K., & Khan, R. (2001). The effects of alliance portfolio characteristics and absorptive capacity on performance. A study of biotechnology firms. *Journal of High Technology Management Research, 12,* 205-226.
Gibbert, M., Ruigrok, W., & Wicki, B. (2008). What passes as a rigorous case study? *Strategic Management Journal, 29,* 1465-1474.
Granovetter, M. (1985). Economic action and social structure: A theory of embeddedness. *American Journal of Sociology, 91,* 481-510.
Gulati, R. (1995). Social structure and alliance formation patterns: A longitudinal analysis. *Administrative Science Quarterly, 40,* 619-652.
Gulati, R. (1998). Alliances and networks. *Strategic Management Journal, 19,* 293-318.
Gulati, R., & Gargiulo, M. (1999). Where do interorganizational networks come from? *American Journal of Sociology, 104,* 1439-1493.
Heimeriks, K. H., Klijn, E. &, Reuer, J. J. (2009). Building capabilities for alliance portfolios. *Long Range Planning, 42,* 96-114.
Hite, J. M., & Hesterly, W. S. (2001). The evolution of firm networks: From emergence to early growth of the firm. *Strategic Management Journal, 22,* 275–286.
Hoffmann, W. H. (2007). Strategies for managing a portfolio of alliances. *Strategic Management Journal, 28,* 827-856.
Holmqvist, M. (2003). A dynamic model of intra- and interorganizational learning. *Organization Studies, 24,* 95-123.
Huber, G. P., & Daft, R. L. (1987). The information environments of organization. In F. Jablin, L. Putnam, K. Roberts, & L. Porter (Eds.), *Handbook of organizational communication* (pp. 130-164). Beverly Hills, CA: Sage.
Iske, P. I. (2010). *Inaugural lecture—Combinatoric innovation. Environments for creation and mobilization of intellectual capital.* Maastricht, Netherlands: Maastricht School of Business and Economics.
Jick, T. D. (1979). Mixing qualitative and quantitative methods: Triangulation in action. *Administrative Science Quarterly, 24,* 602-611.
Jiang, R. J., Tao, Q. T., & Santoro, M. D. (2010). Alliance portfolio diversity and firm performance. *Strategic Management Journal, 31,* 1136-1144.
Khanna, T. (1998). The scope of alliances. *Organization Science, 9,* 340-355.
Koza, M. P., & Lewin, A. Y. (1998). The co-evolution of strategic alliances. *Organization Science, 9,* 795-816.

Lavie, D. (2007). Alliance portfolios and firm performance: A study of value creation and appropriation in the U.S. software industry. *Strategic Management Journal, 28*, 1187-1212.

Lavie, D., & Singh, H. (2011). The evolution of alliance portfolios: The case of Unisys. *Industrial and Corporate Change, 21*, 763-809.

Lorenzoni, G., & Lipparini, A. (1999). The leveraging of interfirm relationships as a distinctive organizational capability: A longitudinal study. *Strategic Management Journal, 20*, 317-338.

March, J. G. (1991). Exploration and exploitation in organizational learning. *Organization Science, 2*, 71-87.

Ozcan, P., & Eisenhardt, K. M. (2009). Origin of alliance portfolios: Entrepreneurs, network strategies, and firm performance. *Academy of Management Journal, 52*, 246-279.

Padula, G. (2008). Enhancing the innovation performance of firms by balancing cohesiveness and bridging ties. *Long Range Planning, 41*, 395-419.

Powell, W. W., Koput, K. W., & Smith-Doerr, L. (1996). Interorganizational collaboration and the locus of innovation: Networks of learning in biotechnology. *Administrative Science Quarterly, 41*, 116-145.

Reuer, J. J., & Ragozzino, R. (2006). Agency hazards and alliance portfolios. *Strategic Management Journal, 27*, 27-43.

Rothaermel, F. T., & Deeds, D. L. (2004). Exploration and exploitation alliances in biotechnology: A system of new product development. *Strategic Management Journal, 25*, 201-221.

Sarkar, M. B., Aulakh, P. S., & Madhok, A. (2009). Process capabilities and value generation in alliance portfolios. *Organization Science, 20*, 583-600.

Schilke O., & Goerzen, A. (2010). Alliance management capability: An investigation of the construct and its measurement. *Journal of Management, 36*, 1192-1219.

Schumpeter, J. A. (1934). *The theory of economic development*. London, UK: Transaction Publishers.

Simonin, B. L. (1997). The importance of collaborative know-how: An empirical test of the learning organization. *Academy of Management Journal, 40*, 1150-1174.

Tjemkes, B. V., Vos, P., & Burgers, K. (2012). *Strategic alliance management*. Abingdon, UK: Routledge.

Volberda, H. W., & Lewin, A. Y. (2003). Co-evolutionary dynamics within and between firms: From evolution to co-evolution. *Journal of Management Studies, 40*, 2111-2136.

Wassmer, U. (2010). Alliance portfolios: A review and research agenda. *Journal of Management, 36*, 141-171.

Yin, R. K. (2003). *Case study research: Design and methods*. Thousand Oaks, CA: Sage.

Yoffie, D., & Slind, M. (2006). *Apple computer, 2006*. Harvard Business School Case no. 3706496.

CHAPTER 7

TO COOPERATE OR NOT TO COOPERATE?

The Dilemma Faced by Inexperienced Firms in R&D Consortia

Isabel Estrada, Natalia Martín-Cruz, and Victor M. Martín-Pérez

ABSTRACT

This chapter examines a phenomenon we observed in a real-life consortium and for which a proper academic explanation is still lacking: under the same challenging conditions, why do only a few inexperienced partners behave cooperatively in the formation process of R&D consortia? Drawing on a multicase study on several consortium partners, we conclude that the reaction of inexperienced firms to the social dilemma inherent to R&D consortia depends on their individual value-creation expectations, which in turn are forged by combinations of incentives of diverse nature. In particular, in the formation of R&D consortia, inexperienced firms are likely to behave cooperatively if they perceive that the consortium entails high strategic potential and brings learning opportunities to realize such potential. At the same time, inexperienced firms facing coopetition tensions are likely to behave noncooperatively, even if they expect important learning and strategic

Strategic Alliances for Innovation and R&D, pp. 159–183

benefits from the consortium. Our evidence, therefore, indicates that the "to cooperate or not to cooperate" dilemma faced by inexperienced partners in multipartner alliances might be better understood as a trade-off between perceived benefits and threats from close interaction with alliance partners.

INTRODUCTION

Research and development (R&D) consortia (i.e., collaborative agreements among multiple partners to jointly conduct a R&D project) bring together many different resources and perspectives and thus afford high potential for synergies and innovation (Doz, Olk, & Ring, 2000). Seeking to seize such potential, public innovation policies across the world are supporting the creation of this particular type of consortia. The Strategic Growth Technologies Program in Denmark, the Multidisciplinary Research Consortia Program in the United Kingdkom, or the National Strategic Consortia for Technical Research Program in Spain can serve as examples. Although large, well-endowed companies have traditionally been the recipients of such public funds, policy makers are increasingly emphasizing the participation of firms with limited experience in R&D collaborative settings, for instance, small firms that operate in low-tech industries. Lack of experience may make the formation of R&D consortia, which is *per se* a complex process (Doz et al., 2000), even more challenging.

R&D consortia, as a particular type of multipartner alliance, entail ambiguous patterns of exchange and reciprocity among the partners (Das & Teng, 2002). Therefore, once a firm has decided to join a R&D consortium, it faces a social dilemma (Zeng & Cheng, 2003): the partner must make a choice between (a) to behave cooperatively and thus contribute to maximizing potential collective gains from the consortium under uncertain conditions of reciprocity *or* (b) not to behave cooperatively, reducing potential collective gains while protecting its own resources from unintended leakages and other partners' opportunism. Such a dilemma is inherently overwhelming for inexperienced firms which, not knowing how to capture collective gains while protecting their own resources (Reuer, Zollo, & Singh, 2002; Sampson, 2005), may find the "not to cooperate" option as the best strategy.

With these basic premises, we started studying the formation process of a government-sponsored consortium formed by inexperienced partners in the Spanish aquaculture industry (i.e., Acuisost Consortium). We observed that different partners, under the same challenging conditions, reacted to the "to cooperate or not to cooperate" dilemma in different ways. In particular, whereas most of the partners displayed noncooperative behaviors, a few partners cooperated actively in pushing the consortium ahead, triggering a cooperative disequilibrium.

Existing research on cooperation in multipartner settings (e.g., Doz et al., 2000; Lavie, Lechner, & Singh, 2007; Thorgren, Wincent, & Eriksson, 2010), while providing important insights, has yet to fully unearth this cooperative phenomenon. Understanding how inexperienced partners react to the R&D consortia social dilemma may bring important implications for theory and for managerial and policy practice as well. Based on the longitudinal study on the Acuisost Consortium, we therefore seek to examine why some inexperienced partners behave cooperatively while others do not during the formation stage of R&D consortia.

We find strong evidence that the reaction of inexperienced partners to the "to cooperate or not to cooperate" dilemma likely depends on their value-creation expectations with the consortium, which in turn are forged by combinations of incentives of diverse nature. In particular, our findings contribute to a richer theoretical understanding on the phenomenon of cooperation in multipartner alliances in two main ways. First, whereas prior research has suggested different but independent effects of several sources of expectations on cooperation, our study offers a more integrative explanation, showing the existence of reinforcing links and tensions between them. In the absence of collaborative experience, a partner is likely to behave cooperatively if it expects strong learning and strategy-related benefits from the consortium and, at the same time, does not perceive significant risk of knowledge leakages inside the consortium. In contrast, inexperienced partners facing coopetition situations in the consortium are likely to behave in a noncooperative fashion: the associated risk of knowledge leakages may even eclipse other learning and strategic expectations, triggering negative reactions to the social dilemma. Second, our study provides some new insights on the relationship between cooperation and reciprocity in the multipartner setting. Relying on social exchange theory (Das & Teng, 2002), scholars have traditionally argued that lack of generalized reciprocity may hinder the sustainability of cooperation in multipartner alliances. Our data nevertheless seem to indicate that, in situations where generalized reciprocity is lacking, direct reciprocity with key actors in the consortium may motivate some inexperienced partners to continue cooperating.

In the following pages of this chapter, we first present the social dilemma that inexperienced firms face when they enter a R&D consortium. After presenting the methodology, we provide the case-based evidence. Subsequently, we discuss the main findings of the study on the Acuisost Consortium. The chapter concludes with the main implications and limitations, as well as some lines for further work.

CONCEPTUAL BACKGROUND

The formation stage of any alliance is critical in the value-creation process: initial conditions are defined (Doz, 1996; Doz et al., 2000) and partners test their compatibility and trustworthiness (Zajac & Olsen, 1993). During this process, cooperation, that is, the relational norm entailing partners' coordinated actions to attain mutual benefits (Anderson & Narus, 1990) facilitates key activities such as the alignment of goals and expectations and the establishment of legitimacy between the partners (Doz et al., 2000; Mahnke & Overby, 2008).

R&D consortia, and multipartner alliances in general, imply multilateral interaction among multiple partners (Lavie et al., 2007). Given these unique logics of collaboration, the formation process of R&D consortia is likely to be particularly complex (Doz et al., 2000) and cooperation herein is likely to prove even more relevant (Hwang & Burgers, 1997; Zeng & Chen, 2003). At the same time, however, achieving cooperation is far from straightforward, since the generalized nature of reciprocity governing exchanges in R&D consortia (Das & Teng, 2002) triggers a social dilemma for partners (Zeng & Chen, 2003). In the following, we briefly describe such a dilemma, eventually explaining how partners are expected to react when they lack prior collaborative experience.

The Social Dilemma in R&D Consortia

According to social exchange theory scholars (Blau, 1964; Das & Teng, 2002), the main difference between multipartner and dyadic alliances resides in the patterns of exchange between partners (and the associated patterns of reciprocity). Dyadic alliances entail direct exchange and reciprocity between two partners, whereas multipartner alliances entail also generalized patterns of exchange and reciprocity among multiple partners (Das & Teng, 2002). In generalized exchanges, partners make individual contributions to the alliance (which could benefit one or more partners), expecting something in return from the alliance (but not necessarily from a specific partner). Reciprocity obligations among the multiple partners are therefore ambiguous (Thorgren et al., 2010).

Given these conditions, partners face a social dilemma in the formation process of multipartner alliances (Zeng & Chen, 2003) which, due to the nature of the exchange resources, may be particularly pronounced in the setting of R&D consortia. Once a firm has decided to enter a multipartner alliance, it must make an important choice: to cooperate or not to cooperate. R&D consortia entail high potential for collective learning and innovation (Doz et al., 2000) but also high levels of relational risk (Das &

Teng, 2000; Sampson, 2005). Partners therefore need to interact closely in order to combine their knowledge and technological resources (Doz et al., 2000; Mothe & Quelin, 2001) and, at the same time, they have to protect such resources from undesirable leakages that may happen through multiple points (Hwang & Burgers, 1997; Oxley & Sampson, 2004). In R&D consortia, both the "to cooperate" and "not to cooperate" positions entail therefore relative advantages and disadvantages.

On the one hand, a partner that behaves cooperatively (i.e., complying with the spirit of cooperation) contributes to making the agreement work and therefore to enhancing the potential collective gains of the alliance (Ariño, 2001; Kumar & Nti, 1998). At the same time, a cooperative partner assumes a vulnerable position in the consortium (Das & Teng, 2002; Thorgren et al., 2010). Since mutual obligations are unclear, cooperating means making blind contributions of resources, not knowing whether, how and by whom such efforts will be reciprocated. On the other hand, a partner that behaves noncooperatively reduces the overall potential of the alliance to generate collective gains but also limits its exposure and vulnerability to partners' opportunism.

How do Partners React to the Social Dilemma in R&D Consortia?

Reviewing the existing literature on strategic alliances, we identify two main groups of studies providing useful insights to address this question. The first group assumes that the degree of cooperation of individual partners depends on their individual value-creation expectations with the exchange relationship (e.g., García-Canal, Valdés-Llaneza, & Ariño, 2003; Mothe & Quelin, 2001; Olk & Young, 1997). Based on this premise, these studies suggest several factors triggering positive or negative reactions to the dilemma in R&D consortia. For example, a number of studies have focused on the phenomenon of coopetition in innovation settings (i.e., firms that compete in the marketplace team up to jointly conduct innovation activities). Collaboration with direct competitors, which is quite common in R&D consortia, may lead to either positive or negative reactions to the social dilemma. On the one hand, competitors often share market interests and have similar knowledge bases, making the creation of mutual benefits possible (Doz et al., 2000), which in turn fosters cooperation. At the same time, however, rivalry among partners may trigger mutual suspicious and mistrust in the consortium (García-Canal et al., 2003; Oxley & Sampson, 2004). In these cases, knowledge sharing and close interaction may be perceived as extremely risky, and partners may thus adopt "not to cooperate" positions to protect themselves vis-à-vis competitors. Other scholars have pointed to the potential of R&D consortia to generate learning and other strategic benefits. Since learning requires frequent and intense interaction and knowledge exchange, strong learning orientation of partners (Hamel, 1991) is likely to promote positive reactions to the dilemma in R&D consortia (e.g. Mothe & Quelin, 2001; Olk & Young, 1997). Likewise, partners are expected to cooperate when the purpose and scope of the R&D consortium are clearly linked to their core strategic activities: the higher the strategic importance of the consortium to the partner, the lower the distance between collective and private gains for that partner (e.g., Lunnan & Haughland, 2008; Mothe & Quelin, 2001).

A second group of studies sheds light on how partners react to the "to cooperate or not to cooperate dilemma" by linking such a dilemma to the alliance experience phenomenon. Existing research has widely demonstrated that firms learn to successfully manage alliances as they accumulate collaborative experience and build collaborative routines (Anand & Khanna, 2000; Simonin, 1997). In R&D consortia, experienced partners, therefore, may be more able to contribute to generating collective gains and to actually capture a greater portion of such gains. In contrast, lack of alliance experience may entail huge challenges for partners (Sampson, 2005). In particular, inexperienced partners lack routines (Zollo & Winter, 2002) to share and, simultaneously, protect valuable resources in R&D consortia. Therefore, existing theory suggests that inexperienced partners may find the "not to cooperate" option as the most rational strategy. After all, it is complex to detect and sanction individual noncooperative behaviors in R&D consortia and partners may feel not particularly guilty for not cooperating (Mahnke & Overby, 2008; Zeng & Chen, 2003).

Both groups of studies have significantly contributed to explaining cooperation dynamics in multipartner settings like R&D consortia. However, we still do not know how different sources of expectations (e.g., coopetition situations, learning orientation and strategic importance) play out when the partner lacks alliance experience and jointly impact the "to cooperate or not to cooperate" decision. This lack of evidence gives meaning to the case study on the Acuisost Consortium.

METHODS

In order to examine why some inexperienced partners behave cooperatively during the formation stage of R&D consortia, we conducted a longitudinal study on the Acuisost Consortium. Our study takes the partner firm as the main unit of analysis and therefore can be described as a "nested" case study (Gibbert, Ruigrok, & Wicki, 2008): focusing on one single setting (i.e., the *Acuisost Consortium*), we study and compare data across multiple cases (i.e., the *partners* of the Acuisost Consortium) that are embedded in this setting. The main advantage of this kind of case study design resides on the opportunity to conduct robust cross-case comparison while minimizing the risk of extraneous variation (Eisenhardt, 1989; Yin, 2003). Below, we present an overview of the research setting and explain how data were collected and analyzed in this study.

Research Setting

The Acuisost Consortium was a R&D consortium aimed to foster sustainable development of Spanish fish aquaculture through four main subprojects (Nutrition, Biosecurity, Waste Management, and Ready Meals). The formation process of the consortium covered the period from January 2006 to March 2009. In January 2006, Aqua-Environ Consulting, a start-up consulting firm, sensed the opportunity to create the Acuisost Consortium (due to confidentiality reasons, the names of the firms used in the chapter are pseudonyms). Spanish aquaculture industry was experiencing a period of great expansion. However, experts were emphasizing the need for innovation efforts to enhance the environmental sustainability of the field. At the same time, sustainability was one of the priorities of the Spanish National Innovation Plan and its several measures, such as the Program of National Strategic Consortia for Technical Research (CENIT Program). Therefore, a collaborative R&D initiative such as the Acuisost Consortium was likely to receive a CENIT grant. Guided by such expectations, Aqua-Environ Consulting started orchestrating the formation of the consortium.

Eventually, 18 firms (i.e., the consortium lead firm and 17 consortium partners) founded the Acuisost Consortium. In CENIT consortia, there are two categories of firms: the lead firm of the consortium (i.e., a partner which assumes coordination and governance responsibilities and holds a higher economic participation in the consortium's budget) and the consortium partners (i.e., firms that play the role of ordinary partners of the consortium). Since participation conditions and expectations of the lead firm and the ordinary partners are inherently different, our analysis focuses on the group of consortium partners. In order to ensure reliable comparison of data across cases, we selected 12 consortium partners following theoretical sampling criteria (Yin, 2003). First, we did not include in the analysis two of the seventeen founding partners that left the Acuisost Consortium during the studied period. Second, according to the criterion of "polar cases" (Eisenhardt, 1989), we concentrated on those partners displaying clear reactions (either positive or negative) to the social dilemma in the Acuisost Consortium. From the fifteen partners that remained involved during the whole studied period, we excluded three partners displaying ambiguous reactions (e.g., behaving cooperatively sometimes and noncooperatively at other times). Table 7.1 displays general information on the selected partners.

Table 7.1. Selected Partners of the Acuisost Consortium

Partner Firms	*Industry*	*Size*	*Age*	*Consortium Subprojects*
BioMilk	Biotechnology	Micro-enterprise	Incumbent	Nutrition
Chemics&Proteins	Chemical	SME	Incumbent	Nutrition
Fish&Meals	Food	SME	Start-up	Nutrition Ready Meals
Green Solutions	Chemical	SME	Incumbent	Biosecurity
Iberian Eels	Aquaculture	SME	Incumbent	Nutrition Wastes
Industrial Packages Co.	Industrial packing	Large	Incumbent	Ready Meals
Mediterranean Aquaculture	Aquaculture	SME	Incumbent	Nutrition Ready Meals
MngProjects	Technology consulting	SME	Incumbent	Nutrition Biosecurity
Northern Trouts Co.	Aquaculture	SME	Incumbent	Biosecurity
Southern Trouts Co.	Aquaculture	SME	Incumbent	Biosecurity Wastes
The Biscuits Co.	Food	Large	Incumbent	Nutrition
WasteMng Co.	Waste management	SME	Start-up	Wastes

Selected partners differed in terms of organizational characteristics such as industry affiliation (e.g., aquaculture, food, biotechnology, chemical industry, etc.), but shared an important feature: all these partners had little prior experience in collaborative R&D (mostly with research centers) and none in R&D consortia. Between January 2007 and March 2009, the partners had to allocate resources, first to structure the consortium and, once public funding was confirmed (in January 2008), to starting up the research project (until March 2009). However, as already mentioned, not all the involved partners contributed equally to the formation of the consortium.

Data Collection and Analysis Procedures

We collected data in the period April 2008 to June 2009, moving from exploratory to more focused actions (Faems, Janssens, Madhok, & Van Looy, 2008; Pettigrew, 1990). This longitudinal approach enhanced the quality of data collection by allowing us to gradually get immersed into the research context and build trust with key informants (Miles & Huberman, 1994), as well as to update our data records on an ongoing basis with real-time information (Langley, 1999). Since the formation process of the Acuisost Consortium covered the period from January 2006 to March 2009, we partially relied on a retrospective approach. Although retrospective approaches are usually associated with efficiency in data collection, reconstruction of past events may be biased (Leonard-Barton, 1990). In order to mitigate these potential problems, we applied several strategies: we conducted structured interviews based on interview protocols, interviewed multiple informants, and triangulated interview data with information from other data sources whenever possible (Faems et al., 2008; Yin, 2003). In all, we conducted 26 interviews (24 face-to-face interviews and 2 phone interviews) with 20 different informants. Interviewees included representatives of (1) the lead firm (CEO, Aquaculture Division Manager, and R&D Manager), (2) Aqua-Environ Consulting (Project Manager), (3) ten partner firms (including six members of the Scientific Committee of the consortium), and (4) five research centers involved in the consortium. In addition, we interviewed the manager of a nonprofit association for innovation in aquaculture, who provided detailed information on the context of the consortium. Furthermore, we collected publicly available information about the consortium, its legal and industrial framework, and its partners (i.e., press releases, policy and corporate reports), analyzed private consortium documents (i.e., consortium reports, consortium agreement, minutes of meetings), collected general survey data on partners (e.g., organizational characteristics, expectations with the consortium), and conducted direct observation in three consortium committees and a consortium annual meeting. Direct observation provided useful data on the partners reactions to the "to cooperate or not to cooperate dilemma" in the studied period.

For example, we observed how Scientific Committee members discussed about knowledge exchange and interaction within the different subprojects of the consortium and collected first-hand information on partners' presence/absence in those consortium meetings.

As common in case study research, we analyzed data on the selected cases through an iterative process, moving from description to explanation (Pentland, 1999) and combining within-case analysis with comparison of data across cases (Eisenhardt, 1989). We started by systematically coding collected data. Based on existing literature, we designed a schema for explicit conversion of raw data from multiple sources into theoretical constructs (Larson, 1993). Emerging findings however triggered some adjustments in the initial schema. For example, we realized that the individual degree of financial involvement in the consortium did not reflect the actual cooperation of all partners, in contrast to what prior studies suggest (e.g. Lavie et al., 2007; Mothe & Quelin, 2001). Given the government-sponsored nature of the Acuisost Consortium, some partners could have economic incentives to increase their participation in the consortium's budget, regardless of their subsequent behavior. This coding process yielded two main outputs, displayed in Table 7.2.

First, based on the two-dimension conceptualization of cooperative behavior (i.e., veracity and commitment) offered by Ariño (2001), we identified two main categories of partners: cooperative and noncooperative partners. Our data clearly indicated a cooperative disequilibrium: only five partners were contributing actively to the formation of the consortium, whereas the involvement of the remaining seven partners in the process turned out to be rather low. Second, the coding process shed light on the main sources of partners' expectations in the consortium, potentially determining their individual reactions to the "to cooperate or not to cooperate dilemma." In particular, we identified those partners facing coopetition in the consortium (e.g., Oxley & Sampson, 2004) and, for each partner, the degree of strategic importance of the consortium (e.g., Mothe & Quelin, 2001) and the level of learning orientation with the consortium (e.g., Hamel, 1991). In several interviews, the R&D Manager of the consortium lead firm, who had detailed information on the research setting and the particular cases, provided insightful feedback on coding process results (Larsson, 1993).

Subsequently, we systematically compared coded data across cases, seeking to identify patterns that could explain why some partners behaved cooperatively during the studied period while others did not. In order to do so, we juxtaposed partners' reactions to the "to cooperate or not to cooperate" dilemma during the studied period against the other variables reflecting sources of expectations (coopetition, learning orientation, and strategic importance). This coding strategy allowed us to reduce all

Table 7.2. Coding Process: Main Outputs

Partner Firms	*Reaction to the Social Dilemma*	*Coopetition*	*Learning Orientation*	*Strategic Importance*
BioMilk	(+) Cooperative behavior	No	High	Medium
Chemics&Proteins	(+) Cooperative behavior	No	High	Medium
Fish&Meals	(-) Noncooperative behavior	No	Low	Medium
Green Solutions	(–) Noncooperative behavior	No	Medium	Low
Iberian Eels	(+) Cooperative behavior	No	High	High
Industrial Packages Co.	(–) Noncooperative behavior	No	Low	Medium
Mediterranean Aquaculture	(–) Noncooperative behavior	Yes	High	High
MngProjects	(–) Noncooperative behavior	No	Low	Medium
Northern Trouts Co.	(–) Noncooperative behavior	Yes	High	High
Southern Trouts Co.	(–) Noncooperative behavior	Yes	Medium	High
The Biscuits Co.	(+) Cooperative behavior	No	High	Medium
WasteMng Co.	(+) Cooperative behavior	No	High	High

the available information to operationalized variables, which in turn facilitated systematic sensemaking and rigorous comparison of data across cases (Eisenhardt, 1991). However, operationalization of case data into formalized constructs may be at the expense of qualitative richness (Dyer & Wilkins, 1991). In order to mitigate this problem, as the last step of the analysis, we came back to insights from individual cases to explain the logic behind the identified patterns (Leonard-Barton, 1990).

REACTIONS TO THE SOCIAL DILMEMMA IN THE ACUISOST CONSORTIUM

In this section, we first provide an overview of the patterns identified from the systematic comparison of data across cases (see Table 7.3).

Subsequently, we provide narrative descriptions of four different cases (two cooperative and two noncooperative partners), focusing on (1) the individual membership conditions of the partners in the consortium, (2) their reactions to the "to cooperate or not to cooperate" dilemma during the studied period, and (3) the impact of the studied sources of expectations on such reactions.

Table 7.3. Positive and Negative Reactions to the Social Dilemma: Identified Patterns

Reaction to the Social Dilemma	*Partner Firms*	*Patterns*
(+) Cooperative behavior	BioMilk The Biscuits Co. Chemics&Proteins Iberian Eels WasteMng Co.	• No coopetition in the consortium • High learning orientation with the consortium • High/medium strategic importance of the consortium
(–) Noncooperative behavior	Mediterranean Aquaculture Northern Trouts Co. Southern Trouts Co.	• Coopetition in the consortium • Medium/high learning orientation with the consortium • High strategic importance of the consortium
	Fish&Meals Green Solutions Industrial Packages Co. MngProjects	• No coopetition in the consortium • Medium/low learning orientation with the consortium • Medium/low strategic importance of the consortium

Since the purpose of these narratives is to illustrate the logic behind the identified patterns, the four cases were selected according to two main criteria (Eisenhardt, 1989). First, in each selected case, the process of interest (i.e., how partners' expectations shaped their reactions to the social dilemma) was transparently observable. Second, each case represents a pattern within its category.

Comparing data across the five cases included in the category of cooperative partners (BioMilk, The Biscuits Co., Chemics&Proteins, Iberian Eels, and WasteMng Co.), we identified a single pattern. None of these partners faced coopetition inside the consortium. Further, they had a high learning orientation with the consortium and attached medium to high levels of strategic importance to the consortium. In contrast, the systematic comparison of the seven noncooperative partners allowed us to identify two different patterns. In the first pattern, the three partners (Mediterranean Aquaculture, Northern Trouts Co., and Southern Trouts Co.) were active in the field of aquaculture and collaborated with direct competitors in the consortium. At the same time, these three partners considered the consortium to be of high strategic importance and held important learning expectations with the consortium. In the second pattern, the four partners (Fish&Meals, Green Solutions, Industrial Packages Co., and MngProjects) were active in industries other than aquaculture and did not face coopetition inside the consortium. They however lacked strong incentives in terms of perceived strategic importance of the Acuisost Consortium and/or expected learning benefits. The following narratives illustrate how the identified patterns led to different reactions to the social dilemma in the Acuisost Consortium.

Positive Reactions to the Social Dilemma in the Acuisost Consortium

Iberian Eels

Iberian Eels, a SME (small and medium enterprise) leading the eel niche market in Spain, was involved in the subprojects of Nutrition and Waste Management, which the firm considered to be critical for its core business areas. The section of the consortium's report devoted to Iberian Eel's strategy illustrates these ideas:

> The strategy of the firm with the consortium entails the development of a new diet for its main product, the eel, that could allow high-quality and more sustainable nutrition and, at the same time, could provide a distinctive sign in the marketplace. (Consortium's report)

In addition, Iberian Eels sought to acquire, during its participation in the consortium, the required knowledge to develop "an integrative system" for managing and transforming organic wastes generated in the day-to-day activities of the company.

Our data indicate that the above perceived strategic and learning benefits led Iberian Eels to be one of the most cooperative partners in the Acuisost Consortium during the period under study. For example, the coordinator of the Waste Management subproject explicitly recognized the strong cooperative efforts of Iberian Eels, which contrasted strongly with the behavior of other partners in the subproject:

> Other firms involved in this subproject are showing their lack of commitment. In contrast, Iberian Eels is playing its role properly ... taking the research project really seriously ... always willing to attend meetings and showing a positive attitude in them. (General Manager of WasteMng)

Similar observations were made about the behavior of Iberian Eels in the Nutrition subproject, where the firm's tasks would be developed in close interaction with the consortium lead firm. As the General Manager of Iberian Eels explained, whereas the Waste Management subproject could offer "higher future potential" than the Nutrition subproject, the technologies to be developed in the latter subproject were part of the commitments made to support the leader in the process of "reinventing aquaculture."

BioMilk

BioMilk, an established biotech microenterprise, perceived important synergies between its core business activities and the aquaculture nutrition field. In particular, the involvement of BioMilk in the Nutrition subproject of the consortium entailed a huge opportunity to explore aquaculture applications of the firm's core products (i.e., probiotics). As stated in the consortium's report, probiotics are "effective tools to inhibit fish pathogens" and, at the same time, "stimulate fish appetite and facilitate digestibility." Collected data also indicated that BioMilk expected to develop new knowledge on both technical and market aspects of the aquaculture nutrition field, since the firm's products had been traditionally applied into human nutrition products:

> New production processes require new technologies and new knowledge ... close collaboration with industry agents may enable the firm to acquire detailed knowledge on the new processes ... to find the most appropriate solutions to apply our products and to get familiar with the new market. (Consortium's report)

As observed in the previous case, strategic and learning expectations motivated BioMilk to be one of the most cooperative partners of the Acuisost Consortium during the studied period. From the very beginning, BioMilk showed its willingness to actively exchange knowledge and to maintain close, frequent, and multilateral interaction with other partners. For example, BioMilk's R&D manager, as a member of the Scientific Committee of the consortium contributed, in the words of the lead firm's R&D Manager, to generating "a miniconsortium within the consortium."

The R&D Manager of BioMilk explicitly complained that some partners were not acting "according to the sharing spirit of CENIT consortia," which was causing "unfair situations." She however emphasized that behaving in such a cooperative way was a win-win strategy with regard to the lead firm of the consortium. BioMilk sought to develop high innovative aquaculture products that could enhance the product portfolio of the lead firm, as a leading nutrition company. In addition, she stressed that the lead firm was providing significant technical support to BioMilk. Developing the knowledge required to explore an important business opportunity in close connection with an important potential client was therefore compensating the cooperative efforts of the firm.

Negative Reactions to the Social Dilemma in the Acuisost Consortium

Mediterranean Aquaculture

Mediterranean Aquaculture, a large fish-farming firm, became involved in the Nutrition and Ready Meals subprojects of the consortium. These two subprojects addressed two of what experts termed as the main challenges of Spanish aquaculture: improving the quality and efficiency of fish nutrition and extending the range of products offered to the market (e.g., introducing new high value-added products like filleted fish). Therefore, the Acuisost Consortium afforded huge strategic opportunities for the development of the core business areas of Mediterranean Aquaculture:

> In recent years, the mass entrance of imported low-cost fish and the reduction in the growth rates of the domestic market are challenging the survival of aquaculture in our country ... the firm seeks to take advantage of the great opportunities offered by the Acuisost Consortium to improve its current competitive position vis-à-vis rivals. (General Manager of Mediterranean Aquaculture)

In addition, Mediterranean Aquaculture conceived its participation in the Acuisost Consortium as an opportunity to develop new technological knowledge. Mediterranean Aquaculture, as many other large firms in the field, had strong production capacities but underdeveloped technological capabilities. For example, our data indicate that the firm lacked specialized staff and lab facilities. The Acuisost Consortium could provide the firm with "the basic technological infrastructure" required to survive in the medium term, since the traditional market of aquaculture was becoming saturated.

Contrary to what might be expected in light of the above observations, Mediterranean Aquaculture did not allocate cooperative efforts during the studied period. Our data point to coopetition tensions as the main cause. In both the Nutrition and Ready Meals subprojects, Mediterranean

Aquaculture collaborated directly with one of its major competitors in the market place. During our interviews, it became clear that the presence of a direct competitor in the consortium led Mediterranean Aquaculture to be more concerned with protecting its core knowledge resources than with actual collaboration, thus preventing the emergence of cooperation:

> We are highly cautious in disseminating relevant information. Let's say that the enemy is at home ... I think this is a mutual feeling. In this subproject we have had no meetings to date and nobody has complained about it. (General Manager of Mediterranean Aquaculture)

Based on his global vision of the consortium, the R&D Manager of the lead firm confirmed that Mediterranean Aquaculture was reluctant to disclose technical aspects such as "the level of progress made in the project, emerging obstacles, and the solutions applied" in order to "prevent competitors from finding out this information."

Fish&Meals

Fish&Meals, a SME producing fish-based prepared meals, participated in the Nutrition and Ready Meals subprojects of the Acuisost Consortium, identifying a twofold opportunity which entailed important strategic implications for the firm:

> On the one hand, exploring the possibilities of revaluing a high-protein by-product from our production process that we still fail to take advantage of.... On the other, developing new ready-to-eat products made from aquaculture fish that could enhance our position in the national market. (Consortium's report)

Despite these expectations, Fish&Meals was one of the least cooperative partners of the Acuisost Consortium during the period studied. For example, Fish&Meals' technical reports were highly generic and did not provide details on the firm's alliance strategy or the technical state of its projects in the consortium, and the firm systematically missed consortium meetings and committees. Our data clearly indicate that the reason for which Fish&Meals lacked motivation to engage in cooperation had to do with its particular view of R&D consortia. The R&D Manager of Fish&Meals described these settings as "shared spaces" in which each partner firm could individually develop its research project, disregarding the importance of close collaboration with other partners to achieve individual objectives. As such, Fish&Meals did not expect to acquire new knowledge from the consortium, neither about technology nor about aquaculture market, that could be essential to capture the above-mentioned opportunities. For example, the R&D Manager of Fish&Meals justified her absences in consortium meetings by arguing that these meetings were actually "a waste of time" and that coordination could be perfectly achieved "by e-mail or phone," neglecting the learning potential from interpartner interaction.

DISCUSSION OF FINDINGS

In this section, we discuss our observations on the Acuisost Consortium and link them to existing literature. We subsequently formulate theoretical propositions explaining why only some inexperienced partners behave cooperatively during the formation stage of R&D consortia.

First, we observed that many of the partners that expected important learning and strategic benefits from the consortium behaved cooperatively during the studied period. Therefore, our evidence supports the findings from prior studies that high perceived levels of learning and strategic potential generate incentives for cooperation. For example, Mothe and Quelin (2001) and Olk and Young (1997) point to strategic importance and learning orientation as two key antecedents, respectively, of resource creation and membership continuity in R&D consortia. At the same time, however, our data on the Acuisost Consortium lead us to deviate from the assumption that the effects of these two factors, when present, are *per se* additive, triggering automatically positive reactions to the "to cooperate or not to cooperate" dilemma. Our comparative analysis suggests that, in the absence of alliance experience, partners that perceive high strategic potential of the consortium are only motivated to cooperate if they expected to acquire from the consortium the knowledge required to realize such potential. For example, informants of BioMilk and The Biscuits Company emphasized that the consortium provided their respective firms with the opportunity to enter into the aquaculture market, which in turn required the acquisition of new knowledge by collaborating closely with other partners of the consortium. In contrast, other noncooperative partners such as Fish&Meals or Industrial Packages Co. perceived significant strategic opportunities from the consortium but neglected the importance of interpartner learning to pursue such opportunities. Therefore, inexperienced partners, even if convinced of the strategic potential of the consortium, may fail to recognize the importance of close interaction and knowledge exchange with other partners, thus lacking motivation to engage in cooperative behavior.

Second, we observed that those partners experiencing coopetition tensions inside the consortium behave noncooperatively during the studied period, whereas none of the cooperative partners faced such tensions. In this regard, our data strongly indicate that coopetition leads inexperienced partners to adopt negative reactions to the social dilemma.

This finding concurs closely with existing research linking competition and instability in alliances (e.g., García-Canal et al., 2003; Park & Russo, 1996). At the same time, our data on the Acuisost Consortium question the idea that competitors are willing to mutually cooperate in pursuit of strategic interests. In particular, we observed that, paraphrasing Park and Russo (1996), *competition eclipsed cooperation* between aquaculture partners. Several informants indicated that aquaculture partners overemphasized the protection of their own knowledge and position vis-à-vis rivals, giving up important strategic opportunities afforded by the Acuisost Consortium, such as solving structural technology problems of Spanish aquaculture or establishing barriers for international competition. Therefore, our data show that, when partners lack alliance experience, perceived risks of knowledge leakages associated to direct competition likely eliminate other cooperative incentives (derived from strategic importance and learning orientation). This evidence parallels competition tensions in intraorganizational settings. Tsai (2002), for example, argues that, in multiunit organizations, competition for obtaining resources between units pursuing common strategic projects discourages knowledge-sharing behaviors. The perception of competitors as "enemies" may be stronger in R&D consortia formed between inexperienced partners than in other coopetition settings described in prior studies (e.g., Gimeno, 2004; Gomes-Casseres, 1996) in which partners, well-endowed with alliance experience, know how to interact in pursuit of common strategic interests without putting at risk their own competitive position.

In sum, factors like learning orientation, strategic importance, and coopetition are important sources of expectations, shaping partners' reactions to the social dilemma in R&D consortia. However, our data strongly indicate that these factors, when considered individually, fail to explain how inexperienced partners react to such a dilemma. In particular, we observed that strategic and learning expectations must be mutually reinforcing to actually prompt inexperienced partners to react positively to the "to cooperate or not to cooperate" dilemma. At the same time, perceived threats of knowledge leakages associated with coopetition are likely to crowd out such cooperative incentives, encouraging inexperienced partners to adopt the "not to cooperate" position. This discussion gives rise to the following propositions:

> **Proposition 1a:** *In the formation of R&D consortia, inexperienced firms are likely to behave cooperatively if they perceive that the consortium (a) entails high strategic potential and (b) brings learning opportunities to realize such potential.*

Proposition 1b: *In the formation of R&D consortia, inexperienced firms facing coopetition tensions are likely to behave noncooperatively, even if they expect important learning and strategic benefits from the consortium.*

The above discussion raises the question of whether the above combination of learning and strategic benefits, when coopetition tensions are lacking, fully explains how cooperative partners frame and react to the social dilemma. Put differently, which role did the reciprocity-cooperation link play in the Acuisost Consortium? Social exchange scholars have suggested that reciprocity sustains cooperation over time (Anderson & Narus, 1990; Lado, Boyd, & Hanlon, 1997). In contrast, when generalized reciprocity is lacking in a multipartner alliance (as occurred in the Acuisost Consortium), incentives for cooperation may disappear (Das & Teng, 2002; Zeng & Chen, 2003). Our observations seem to suggest that bilateral or direct reciprocity with key actors in the consortium (lead firm), as long as coopetition tensions are lacking, may be strong enough to sustain cooperation during the formation stage. In the Acuisost Consortium, a sense of direct reciprocity emerged between some of the cooperative partners (those working on new aquaculture nutrition products) and the lead firm of the consortium (a leading company in aquaculture nutrition). Informants of these partners complained about "the imbalance in partners' contributions" but emphasized that the support they were receiving from the lead firm, with which they envisioned future commercial agreements, compensated for their cooperative efforts.

CONCLUSION AND IMPLICATIONS

Based on a longitudinal multiple case study, we have examined why only some inexperienced partners behave cooperatively during the formation stage of R&D consortia. Below, we discuss the main implications of the study, for research and practice, as well as the main limitations and some interesting ideas for future work.

Implications for Research

Whereas the positive effects of strong alliance experience have been widely discussed (e.g., Sampson, 2005; Zollo & Winter, 2002), existing alliance research rarely looks at firms lacking collaborative experience. Our study focuses on the social dilemma of multipartner alliances for partners that do lack collaborative experience. In doing so, we present an integrative view on the factors shaping partners' expectations and their impact on the reactions to the "to cooperate or not to cooperate" dilemma. Existing studies seem to suggest that factors like coopetition (García-Canal et al., 2003), learning orientation (Hamel, 1991), or strategic importance (Mothe & Quelin, 2001) trigger different but somewhat independent effects on cooperation. Our data, however, indicate that the effects of these factors are intertwined. In R&D consortia created by inexperienced partners, positive reactions to the social dilemma are likely associated with two simultaneous conditions: (a) partners perceive that the consortium entails high strategic potential and brings learning opportunities to realize such potential, and (b) partners do not perceive threats of unintended knowledge spillovers inside the consortium. In contrast, coopetition situations among inexperienced partners are likely to trigger negative reactions to the social dilemma. Under situations of coopetition, inexperienced partners are likely to perceive that potential losses from knowledge leakages outweigh potential learning and strategic benefits.

Overall our findings provide a more comprehensive understanding on how inexperienced partners frame and react to the overwhelming dilemma of multipartner collaborative settings such as R&D consortia (Zeng & Chen, 2003). By identifying interactions between several antecedents of alliance behavior, our study shows that the reactions of partners to the social dilemma are contingent on their individual value-creation expectations with the consortium, which in turn reflect a trade-off between the expected benefits and costs from interaction with alliance partners. Our observations suggest therefore an analogy between the social dilemma of "to cooperate or not to cooperate" faced by inexperienced partners in R&D consortia and the broader strategic dilemma of "to invest or not to invest" in new projects under uncertainty (McGrath & Nerkar, 2004). In addition, we offer some novel insights on the cooperation-reciprocity relationship in such complex collaborative settings. Our findings seem to indicate that cooperation may be possible in specific parts of the consortium where direct reciprocity prevails, even when partners perceive a lack of generalized reciprocity in the consortium that would make it difficult for members to cooperate in a sustained manner (Das & Teng, 2002; Thorgren et al., 2010). In particular, perceived bilateral reciprocity with key actors in the consortium could be strong enough to motivate some partners with high value-creation expectations to continue cooperating.

Implications for Managerial and Policy Practice

Based on the case study on the Acuisost Consortium, we offer some recommendations for the management of R&D consortia formed by inexperienced partners and for public policies funding this kind of collaborative initiatives. First, in order to promote positive reactions to the social dilemma, selection of potential partners must take into account not only theoretical synergies that may emerge among the partners, but also the compatibility of their multiple value-creation expectations. Furthermore, when approaching potential partners, triggering entities must convey the spirit of multilateral cooperation and emphasize that the potential strategic benefits the consortium affords requires interactive learning, making clear the commitments that partners will be expected to undertake in this regard. In addition, establishing several subprojects may improve efficiency and coordination in R&D consortia (Dyer & Nobeoka, 2000), but the composition of each subproject should be carefully defined. In particular, our findings suggest preventing coopetition tensions in order to support knowledge sharing within each consortium subproject.

Public funds may bring otherwise unattainable R&D opportunities to inexperienced firms, often SMEs in non-high-tech industries. Many informants of the Acuisost Consortium emphasized this point during interviews. However, public policies should help those firms to learn how to cooperate in such complex settings. For example, government agencies may act as triggering entities of microconstellations formed by competitors in industries like Spanish aquaculture facing global competitive challenges. As learning is a matter of experience accumulation, these initiatives may favor the long-term performance of macroconstellations such as R&D consortia. Likewise, in deciding allocation of funds, public agencies should assess the relational potential of each applicant consortium. Based on our observations, we suggest paying special attention to (a) whether partners are convinced that the consortium could bring important opportunities for strategic development and, at the same time, the required knowledge to seize such opportunities, and (b) whether coopetition tensions could make even more complex the formation process of the consortium by hindering the emergence of cooperation.

Limitations and Future Research

While this study provides interesting insights, some important limitations should be noted. Taken together, these insights and limitations suggest some promising avenues for future research. First, the reduced risk of extraneous variation inherent to "nested" case study approaches allows rigorous analytical generalization of findings (Eisenhardt, 1989), but may reduce theoretical richness (Gibbert et al., 2008). Our findings are clearly contextualized in the setting of the Acuisost Consortium and thus prevent broader replication. Future research should explore how inexperienced partners react to the social dilemma in other R&D consortia, accounting for different industrial and institutional settings. Furthermore, in order to ensure rigor in comparative analysis, we did not study some partners displaying ambiguous reactions to the "to cooperate or not to cooperate" dilemma. To some extent, our framework offers a simplistic picture focused on what existing research (e.g., Ariño, 2001; Kumar & Nti, 1998) considers the typical modes of alliance behavior (cooperative and noncooperative behavior). Examining why some partners adopt *atypical* modes of alliance behavior in R&D consortia could be an interesting line for further work.

In addition, our findings raise the question of under which conditions inexperienced partners start building their alliance capabilities (De Man, Duysters, & Saebi, 2010). Cooperation entails investments of resources that, together with other initiatives, may lead to the creation of collaborative routines (Dyer & Singh, 1998). Cooperative behavior may thus reflect part of the initial alliance capability-building investments of inexperienced firms. Given the relational dimension of alliance capabilities (e.g., Sarkar, Aulakh, & Madhok, 2009), the decision of a firm to start creating alliance capabilities or not to do so is likely related to its reciprocity perceptions. For example, regarding partners' reciprocity perceptions, our findings indicate a potential tension between the presence of coopetition and the presence of a strong relationship with the consortium leader.

Finally, the Acuisost Consortium case provides indications that key actors in alliances could use governance mechanisms to foster cooperation, for example by reinforcing partners' initial cooperative frames (Dyer & Singh, 1998; Sarkar et al., 2009). Yet much remains unknown about the role played by lead firms and/or triggering entities in R&D consortia from a relational governance perspective (Doz et al., 2000; Faems et al., 2008). Further exploration of these questions could contribute to a better understanding of the unique dynamics of cooperation in multipartner collaborative settings.

ACKNOWLEDGMENTS

The study reported in this chapter is part of a broader research project on the Acuisost Consortium, which received financial support from the European Social Fund and the Junta de Castilla y León [Contract for Novel Researchers 2008-2012], from the Ministerio de Economía y Competitividad (Plan Nacional de I+D+i) of Spain [ECO2012-32075], and the 10th Juan José Renau Piqueras Award in Economic Research of the Vall d'Albaida University Foundation (University of Valencia).

REFERENCES

Anand, B. N., & Khanna, T. (2000). Do firms learn to create value? The case of alliances. *Strategic Management Journal, 21*, 295-315.

Anderson, J. C., & Narus, J. A. (1990). A model of distributor firm and manufacturer firm working partnerships. *Journal of Marketing, 54*(1), 42-58.

Ariño, A. (2001). To do or not to do? Noncooperative behavior by commission and omission in interfirm ventures. *Group & Organization Management, 26*(1), 4-23.

Blau, P. M. (1964). *Exchange and power in social life*. New York, NY: Wiley.

Das, T. K., & Teng, B. (2000). Instabilities of strategic alliances: An internal tensions perspective. *Organization Science, 11*, 77-101.

Das, T. K., & Teng, B. (2002). Alliance constellations: A social exchange perspective. *Academy of Management Review, 27*, 445-456.

De Man, A. P., Duysters, G., & Saebi, T. (2010). Alliance capability as an emergent theme: Past, present, future. In T. K. Das (Ed.), *Researching strategic alliances: Emerging perspectives*. Charlotte, NC: Information Age Publishing.

Doz, Y. L. (1996). The evolution of cooperation in strategic alliances: Initial conditions or learning processes? *Strategic Management Journal, 17*(Summer special issue), 55-83.

Doz, Y. L., Olk, P. M., & Ring, P. S. (2000). Formation processes of R&D consortia: Which path to take? Where does it lead? *Strategic Management Journal, 21*, 239-266.

Dyer, J. H., & Singh, H. (1998). The relational view: Cooperative strategy and sources of interorganizational competitive advantage. *Academy of Management Review, 23*, 660-679.

Dyer, J. H., & Nobeoka, K. (2000). Creating and managing a high-performance knowledge-sharing network: The Toyota case. *Strategic Management Journal, 21*, 345-367.

Dyer, W. G., & Wilkins, A. L. (1991). Better stories, not better constructs, to generate better theory: A rejoinder to Eisenhardt. *Academy of Management Review, 16*, 613-619.

Eisenhardt, K. M. (1989). Building theories from case study research. *Academy of Management Review, 14*, 532-550.

Eisenhardt, K. M. (1991). Better stories and better constructs: The case for rigor and comparative logic. *Academy of Management Review, 16*, 620-627.

Faems, D., Janssens, M., Madhok, A., & Van Looy, B. (2008). Toward an integrative perspective on alliance governance: Connecting contract design, trust dynamics, and contract application. *Academy of Management Journal, 51*, 1053-1078.

García-Canal, E., Valdés-Llaneza, A., & Ariño, A. (2003). Effectiveness of dyadic and multi-party joint ventures. *Organization Studies, 24*, 743-770.

Gibbert, M., Ruigrok, W., & Wicki, B. (2008). What passes as a rigorous case study? *Strategic Management Journal, 29*, 1465-1474.

Gimeno, J. (2004). Competition within and between networks: The contingent effect of competitive embeddedness on alliance formation. *Academy of Management Journal 47*, 820-842.

Gomes-Casseres, B. (1996). *The alliance revolution: The new shape of business rivalry*. Cambridge, MA: Harvard University Press.

Hamel, G. (1991). Competition for competence and interpartner learning within international strategic alliances. *Strategic Management Journal, 12*(Special issue on global strategy), 83-103.

Hwang, P., & Burgers, W. P. (1997). The many faces of multi-firm alliances: Lessons for managers. *California Management Review, 39*(3), 101-117.

Kumar, R., & Nti, K. O. (1998). Differential learning and interaction in alliance dynamics: A process and outcome discrepancy model. *Organization Science, 9*, 356-367.

Lado, A. A., Boyd, N. G., & Hanlon, S. C. (1997). Competition, cooperation, and the search for economic rents: A syncretic model. *Academy of Management Review, 22*, 110-141.

Langley, A. (1999). Strategies for theorizing from process data. *Academy of Management Review, 24*, 691-710.

Larsson, R. (1993). Case survey methodology: Quantitative analysis of patterns across case studies. *Academy of Management Journal, 36*, 1515-1546.

Lavie, D., Lechner, C., & Singh, H. (2007). The performance implications of timing of entry and involvement in multipartner alliances. *Academy of Management Journal, 50*, 578-604.

Leonard-Barton, D. (1990). A dual methodology for case studies: Synergistic use of a longitudinal single site with replicated multiple sites. *Organization Science, 1*, 248-266.

Lunnan, R., & Haugland, S. A. (2008). Predicting and measuring alliance performance: A multidimensional analysis. *Strategic Management Journal, 29*, 545-556.

Mahnke, V., & Overby, M. L. (2008). Failure sources in R&D consortia: The case of mobile service development. *International Journal of Technology Management, 44*(1), 160-178.

McGrath, R., & Nerkar, A. (2004). Real options reasoning and a new look at the R&D investment strategies of pharmaceutical firms. *Strategic Management Journal, 25*, 1-21.

Miles, M. B., & Huberman, A. M. (1994). *An expanded sourcebook of qualitative data analysis*. Thousand Oaks, CA: Sage.

Mothe, C., & Quelin, B. V. (2001). Resource creation and partnership in R&D consortia. *Journal of High Technology Management Research, 12*(1), 113-138.

Olk, P., & Young, C. (1997). Why members stay in or leave an R&D consortium: Performance and conditions of membership as determinants of continuity. *Strategic Management Journal, 18*, 855-877.

Oxley, J. E., & Sampson, R. C. (2004). The scope and governance of international R&D alliances. *Strategic Management Journal, 25*, 723-749.

Park, S. H., & Russo, M. V. (1996). When competition eclipses cooperation: An event history analysis of joint venture failure. *Management Science, 42*, 875-890.

Pentland, B. T. (1999). Building process theory with narrative: From description to explanation. *Academy of Management Review, 24*, 711-724.

Pettigrew, A. M. (1990). Longitudinal field research on change: Theory and practice. *Organization Science, 1*, 267-292.

Reuer, J. J., Zollo, M., & Singh, H. (2002). Post-formation dynamics in strategic alliances. *Strategic Management Journal, 23*, 135-151.

Sampson, R. C. (2005). Experience effects and collaborative returns in R&D alliances. *Strategic Management Journal, 26*, 1009-1031.

Sarkar, M. B., Aulakh, P. S., & Madhok, A. (2009). Process capabilities and value generation in alliance portfolios. *Organization Science, 20*, 583-600.

Simonin, B. L. (1997). The importance of collaborative know-how: An empirical test of the learning organization. *Academy of Management Journal, 40*, 1150-1174.

Thorgren, S., Wincent, J., & Eriksson, J. (2010). Too small or too large to trust your partners in multipartner alliances? The role of effort in initiating generalized exchanges. *Scandinavian Journal of Management, 27*, 99-112.

Tsai, W. (2002). Social structure of "coopetition" within a multiunit organization: Coordination, competition, and intraorganizational knowledge sharing. *Organization Science, 13*, 179-190.

Yin, R. K. (2003). *Case study research, design and methods*. Thousand Oaks, CA: Sage.

Zajac, E. J., & Olsen, C. P. (1993). From transaction cost to transactional value analysis: Implications for the study of interorganizational strategies. *Journal of Management Studies, 30*, 131-145.

Zeng, M., & Chen, X. P. (2003). Achieving cooperation in multiparty alliances: A social dilemma approach to partnership management. *Academy of Management Review, 28*, 587-605.

Zollo, M., & Winter, S. G. (2002). Deliberate learning and the evolution of dynamic capabilities. *Organization Science, 13*, 339-351.

CHAPTER 8

FROM CONTINGENCIES TO PARADOXES

Exploring Tensions in Research on Alliance Portfolios and Innovation

Leonardo Corbo, Jing Sun, and Weilei (Stone) Shi

ABSTRACT

Evidence from previous research shows that our understanding of the relationship between alliance portfolios and innovation is still limited. More recently, scholars have focused on the characteristics of alliance portfolios to provide finer-grained results. Despite the importance of alliance portfolios for innovation, results remain somewhat mixed leading to tensions. To address these tensions, researchers have explored contingency-based arguments to assess the conditions under which alliance portfolios can increase or decrease innovative performance. In this chapter we offer an alternative approach using paradox to reconcile conflicting views. Our propositions suggest that treating dualities as complements rather than substitutes promises to enrich the discourse on the alliance portfolio—innovation relationship.

Strategic Alliances for Innovation and R&D, pp. 185–208

INTRODUCTION

An increasing number of firms have engaged in multiple alliance initiatives over the past decades. This makes corporate performance hinge increasingly on the effective management of alliance portfolios (Kale & Singh, 2009). Our understanding of the performance implications, particularly innovation performance, of these alliance portfolios has been quite limited. The emergence of alliance portfolio shifts scholarly attention from single dyads to the overall firm alliance programs (e.g., Hagedoorn & Schakenraad, 1994). This allows researchers the opportunity to study the potential synergies and tradeoffs companies experience when they pursue multiple alliances.

We define a portfolio, also known as "egocentric network," as a firm's set of direct ties (Das & Teng, 2002). The characteristics of the portfolio of alliances a firm maintains can determine the extent to which it is capable of appropriating value deriving from innovation. Partner characteristics have a strong influence on whether and how well the firms engaging in an alliance learn from each other. Within this research domain, researchers have often characterized alliances as wellsprings of innovation and new capabilities (Hamel, 1991; Leonard-Barton, 1995). Many studies have shown that alliance portfolios in which firms are embedded can enhance firm learning and innovation (Shan, Walker, & Kogut, 1994; Soh, 2003). Despite the importance of alliance portfolios for innovation, results remain somewhat mixed leading to tensions such as centrality (Powell, Koput, & Smith-Doerr, 1996) versus structural holes (Burt, 1992), direct ties versus indirect ties (Ahuja, 2000a), and diversity versus homogeneity (Goerzen & Beamish, 2006), to name a few.

In this chapter, we take a paradox approach, which goes beyond the traditional contingency-based arguments. A paradox approach recognizes the duality within firms' overall alliance initiatives. Rather than stating conditions under which the relationship holds (the contingency way), the paradox approach focuses on how organizations can attend to competing demands and meet divergent goals simultaneously (Andriopoulos & Lewis, 2009). A paradox is an idea involving two opposing thoughts or propositions which, however contradictory, are equally necessary to convey a more imposing, illuminating, life-related or provocative insight into truth than either factor can muster in its own right (Slaatte, 1968, p. 4). In the context of our study, this refers to how firms can effectively deal with the tension inherited in their alliance portfolios. To further articulate these tensions, we focus on three different levels of analysis, namely firm, dyadic alliance, and alliance portfolio. To manage these tensions successfully, we argue that firms should attend to these different levels from a holistic perspective since each level is highly embedded and interrelated with other levels. We also identify three different tensions through literature review, that is, (1) trust versus opportunism, (2) complementarity versus similarity, and (3) strong versus weak ties. We focus on these tensions since they are the most important and relevant ones across prior studies in facilitating and constraining innovation performance.

Our study makes the following contribution to the alliance portfolio and paradox literature. We view tensions as inherited within the alliance initiatives. Different conflicting forces are complementary rather than contradictory with each other. To this end, it is important to incorporate such tensions into theory building and offer effective implementation of solutions to the paradox. Our study is resonant with the call for an integrative understanding among the solutions to organizational paradox (Boumgarden, Nichkerson, & Zenger, 2012; Jansen, Tempelaar, van den Bosch, & Volberda, 2009; Raisch & Birkinshaw, 2008; Raisch, Birkinshaw, Probst, & Tushman, 2009; Smith & Lewis, 2011). Doing so, as Smith and Lewis (2011, p. 392) pointed out, "involves consistent inconsistency as managers frequently and dynamically shift decisions." Therefore, asymmetry and tension should be viewed as part of the long-term contradiction over time when firms manage multiple alliance initiatives across different system states.

ALLIANCE PORTFOLIOS, INNOVATION AND COMPETING LOGICS

Innovation has been recognized as playing a key role in the long-term survival of firms and it has been a topic of continuous investigation by management scholars. It allows firms to adapt to environmental dynamism and to obtain competitive advantage by providing new products or services, by decreasing the cost of existing products or services, by improving their quality, and by delivering better business models (Hauser, Tellis, & Griffin, 2006; Huston & Sakkab, 2006). To be innovative, firms need to search constantly for new ideas to foster creativity and build up knowledge internally to develop new products or services. Moreover, as it is uncertain which technology will be successful, firms must invest in a variety of technologies. Few firms are able to do that by themselves. This suggests that the "do not go it alone" strategy could be a preferred option to innovate especially in knowledge-intensive industries. To get access to really new ideas, external partners are often a more helpful and less costly solution compared to increasing internal research and development (R&D) activities.

In recent years, there has been unprecedented growth in strategic alliances and other forms of collaboration especially in technology-intensive industries such as biotechnology, software, aerospace, and semiconductors, to name a few. Alliances can give high technology firms access to valuable resources, in some cases unattainable otherwise, such as new knowledge, patents, licenses, and talents, but also to new markets and new customers. For instance, in the semiconductor industry many of the horizontal alliances are created with the intent to acquire device or manufacturing technology from strategic partners (Stuart, 2000). Given these premises, it seems logical to argue that in terms of innovative performance, alliances are beneficial to firms joining them. Yet, alliance failure rates remain pretty high with at least one out of two alliances not reaching their initial goals (Dyer, Kale, & Singh, 2001).

There are different motives for firms to join alliances. When it comes to innovation strategy, interorganizational collaboration might imply access to complementary assets needed to turn innovation projects into a commercial success (Hagedoorn, 1993). This is, for instance, very frequent in alliances between small biotechnology firms and large pharmaceuticals where the former have innovative resources and capabilities but lack marketing ones which are provided by the latter. Interfirm alliances may also help partners sharing the costs of research and development (Hagedoorn, 2002) as well as encouraging the transfer of codified and tacit knowledge (Doz & Hamel, 1998). Another reason why alliances may increase innovativeness is that they serve as radars (Duysters & de Man, 2003). Alliances enable firms to scan their environment for promising new technologies while keeping costs low. At the end of the scan process, less promising technologies will be abandoned while the most promising ones may be introduced into the company's portfolio.

Besides the aforementioned benefits, alliances may also impact innovation negatively because knowledge transfer across organizations is notoriously difficult (de Man & Duysters, 2005). Differences in corporate culture, processes and knowledge base may impede a smooth transmission of knowledge between partners (Lane & Lubatkin, 1998). Furthermore, it may happen that a firm joins an alliance with the objective to learn as much and as quickly as possible from the partner before deciding to leave the alliance. Alliance termination is not just limited to underperforming alliances. High-performing alliances may be terminated early due to conflicting interests, changed priorities, quickly accomplishing their intended goals, or one of the partners has won the learning race (Larsson, Bengtsson, Henriksson, & Sparks, 1998, p. 287).[1]

As mentioned above, firms basically enter multiple alliance agreements in order to overcome uncertainty and optimize risk, while increasing their economic returns through the acquisition of necessary resources. This suggests the need for firms to focus less on the single dyads and more of their alliance portfolio. Companies may be able to appropriate more value from their alliances when they manage them as a portfolio. Although a single tie can be useful, a firm's portfolio of ties is likely to be more crucial to the firm's performance, thus placing portfolios at the heart of strategic interest (Ozcan & Eisenhardt, 2009). A firm's alliance portfolio serves as a more precise signal to the market on a firm's capabilities and performance than the information available from a single alliance. Using the alliance portfolio as a unit of analysis instead of focusing on the individual alliances, researchers have explored, among others, how the number of alliances affect firms' innovation output (Ahuja, 2000a), new product development (Baum, Calabrese, & Silverman, 2000), and revenue growth (Rothaermel, 2001). A number of recent contributions have argued that by adopting a portfolio approach to external innovation information sources, firms put themselves in a better position to achieve and sustain innovation (e.g., Katila & Ahuja, 2002).

More recently, studies on alliance portfolios have focused on their attributes and characteristics to explain the reasons behind high-performing alliance portfolios. An important distinction to be made here is that between the characteristics of the focal firm and those of the focal firms' partners. Firm-level attributes include size, experience as well as network characteristics (e.g., centrality). Partner-level attributes that have been examined include reputation and quality, resource contributions and internationality of the partners. Both play an important role in determining the sign (i.e., positive or negative) and the direction (i.e. linear or curvilinear) of the alliance portfolio—innovative performance relationship. For instance, Gulati and Higgins (2003) found that alliance portfolios did not matter for entrepreneurial firms (a firm-level attribute) in the biotech sector at the time of the firms' initial public offerings (IPOs) and that a young biotech firm's strategic alliances with well-established pharmaceutical firms (a partner-level attribute) did not matter to IPO success, especially under cold market conditions for new equity offerings. The underlying rationale for focusing on the characteristics is to provide a more nuanced explanation of the relationship between alliance portfolios and innovative performance. Some studies have embraced this view by showing that the advantage of an alliance portfolio is determined not so much by the number of ties a firm maintains but by the characteristics of the firms that a focal organization is connected to (Stuart, 2000). Put differently, adding more ties to a firm's portfolio does not automatically imply that there will be a positive effect on a firm's innovative output. Instead, there are cases in which increasing the number of direct ties can be counterproductive. For instance, a large alliance portfolio creates a risk of dealing with many unfamiliar streams of knowledge that are progressively difficult to integrate (Ahuja & Katila, 2004). More partners implies more risk of opportunism and arbitrage which increase complexity as well as the time and effort the firm needs to dedicate to monitor its partners. Accordingly, marginal benefits of additional alliances are expected to decrease, whereas marginal costs are expected to increase (Ahuja, 2000a). Attributes of alliance portfolios such as the structure of the alliances (horizontal versus vertical alliances), the nature of alliances (explorative versus exploitative alliances), diversity, and complexity have all proved to influence the innovative performance of companies. For example, Baum and his colleagues (2000) found that biotech ventures with portfolios containing many ties to diverse partners were more likely to survive. All in all, this and other studies seem to indicate the need to incorporate firm, dyadic and portfolio level characteristics in the study of alliance portfolios.

FROM A CONTINGENCY APPROACH TO A PARADOX PERSPECTIVE

Research on alliance portfolios has grown only recently. The still-limited amount of studies focusing on the consequences of alliance portfolios are even more limited if we select those that deal with innovative and economic performance. The generalizability of the existing studies is somewhat limited, because the majority of them have focused on a limited number of industries (often biotech and pharmaceutical) and/or a particular type of firms such as start-ups and entrepreneurial firms (e.g., Shan et al., 1994). Moreover, findings of how alliance portfolios impact innovation often provide contradictory explanations. To provide an example, whereas Stuart, Hoang, and Hybels (1999) found that young firms with prominent alliance partners completed their IPO faster and earned greater valuations from the IPO, Gulati and Higgins (2003) showed that for young firms, forming alliances with well-reputed partners did not matter to IPO success. Despite those limits, conflicting results should not discourage researchers from developing new predictions or testing the previous ones in other settings. Indeed, the dynamic tensions of juxtaposed opposites (i.e., paradox) (Rosen, 1994) underpin much organization and management scholarship and practice. Organizing itself raises multiple tensions.

Instead of proposing a one best way, researchers have progressively moved to offer alternative solutions, mainly based on contingency theory. Since the 1960s contingency theorists have focused on the conditions under which alternative elements of tensions are most effective (e.g., Lawrence & Lorsch, 1967). According to this lens, organizational systems are successful when they achieve alignment or fit among internal elements and with the external environment. The role of management is to recognize and then resolve tensions. In the context of the alliance portfolio—innovation relationship, motivated by the observation that not all firms benefit equally in terms of their ability to innovate from their alliance portfolios, several management scholars have recently advocated and embraced a contingency perspective (e.g., Schilling & Phelps, 2007). This requires a shift in focus, from a main-effects perspective toward a contingency one to address questions such as why some firms benefit more in terms of their ability to introduce new products on the market (i.e., innovation) from certain alliance portfolio configurations than others. For instance, Shan and his colleagues (1994) found that the relationship between a start-up's alliance portfolio and its innovative output is moderated by the size of its alliance partners. Overall, previous studies introducing contingency-based arguments have concluded that the extent to which a portfolio of alliances influences the innovativeness of firms depends on certain conditions such as similar knowledge backgrounds of the partners, a higher level of alliance capability and more intense relationships among others.

An alternative approach is offered by the paradox perspective. As with contingency theory, a paradox perspective explores tensions across phenomena and levels, but in contrast to contingency theory, a paradox perspective assumes that tensions persist within complex and dynamic systems (Smith & Lewis, 2011). In other words, paradox studies explore how organizations can attend to competing demands simultaneously. The management of this duality hinges on exploring the tension in a creative way that captures both extremes. Thus, managing paradox emerges as an opportunity to explore tensions at the boundary level that reveal themselves in terms of mixed messages and contradictions. In the context of alliances and innovation, innovation strategies are translated into a number of projects addressing both short and medium term or even longer term innovation objectives. At the same time, it is noticeable that firms trying to achieve a broad spectrum of innovation activities are faced with multiple, often opposing, demands that confront them with the challenge of reconciling conflicting requirements. Efforts to reconcile tensions have been usually isolated studies (e.g., Andriopoulos & Lewis, 2009) that have not received systematic attention yet.

Previous studies on alliance portfolios and innovation have generally focused on three different levels where also dualities have emerged. In more detail, attributes can be defined at the firm, the dyadic and the portfolio level. The first set of attributes defines how the focal firm's characteristics relative to its portfolio affect innovation outputs. Examples include the number of alliances the firm has (i.e., portfolio size), a firm's absorptive capacity (i.e., the ability to value and apply knowledge) as well as the firm's experience in managing partnerships (i.e., the presence of an alliance function within the firm). We then identify a relational dimension which mainly focuses on the dyadic ties within the alliance portfolio. In other words, the focus here is on the quality of the ties that is shifting attention to what is exchanged between partners (i.e., the content of ties). Lastly, another important aspect of alliance portfolios is the set of alliance partners to which a firm is connected. There is broad evidence that partner characteristics influence the firm's ability to innovate (e.g., Gulati & Higgins, 2003). Examples include reputation and quality of the partners, resource contributions, and internationality of the partners. This stream of research adheres to the notion that alliance portfolios provide the focal firm with access to network resources (Gulati, 1999) that extend its opportunity set and potentially enhance its innovative performance. For the purposes of our study we identify three main tensions, respectively at the firm, the alliance and the portfolio levels emerging in past research on alliance portfolios and innovation, namely: (1) trust versus opportunism, (2) complementarity versus similarity, and (3) strong versus weak ties.

TRUST VERSUS OPPORTUNISM

Viewing innovation as recombination of knowledge (Fleming, 2001), trust is conceived as an important catalyst to flourish knowledge transfer and innovation (Cowan, Jonard, & Zimmermann, 2007; Dhanaraj & Parkhe, 2006). Contrarily, opportunism is a central concern in the economics of innovation (Arrow, 1974; , 1942; Teece, 2000) and has been identified as a major hurdle in effective knowledge flows for innovation (Nieto & Santamaria, 2007; Perks & Jeffery, 2006). However, we have limited knowledge on how the tension between trust and opportunism impacts on innovation performance, especially in the context of alliance portfolios. Therefore, it may be possible to argue that trust and opportunism are two fundamental elements in managing alliance portfolios to generate innovation. Studying the tension between these two elements can provide a more profound insight on the relationship between alliance portfolios and innovation.

In innovational alliance portfolios, trust plays an important role in minimizing the threat of opportunism (Segrestin, 2005). It generates economic rents and contributes to innovation performance by reducing uncertainty, diminishing transaction costs, strengthening social embeddedness, diffusing information and increasing resource exchange (Soh, Mahmood, & Mitchell, 2004; Tsai & Ghoshal, 1998). First, given the high uncertainty characterizing the collaborative innovation process, a high level of trust is essential for network orchestration (Dhanaraj & Parkhe, 2006). The relational trait of trust discourages free riding and trust is self-reinforced through social norms development (Ostrom, Walker, & Gardner, 1992). Intuitively, trust creates cognitive expectations and moral guidance for interactions among alliance partners. Moreover, trust brings social norms and it is strengthened via repeated collaborations (Mesquita

& Lazzarini, 2008). Trust serves also as an alternative mechanism to formal governance structure to safeguard against partner's opportunistic behaviors (Quintana-Garcia & Benavides-Velasco, 2004). Especially, when formal contracts are costly to sign and the institutional environment is not supportive, trust is a crucial factor for engaging in uncertain innovational activities (Mesquita & Lazzarini, 2008). With higher trust, firms are able to negotiate lenient and less formal alliance governance structures, which enhances flexibility in their collaborative innovation activities. On the contrary, employing explicit knowledge protective policies and procedures may escalate distrust because it triggers partner's attention to opportunism (Nielsen & Nielsen, 2009).

Second, transferring tacit knowledge is crucial for being innovative. Tacit knowledge usually resides in relational ties where two alliance parties have high trust toward each other (Nooteboom,Vanhaverbeke, Duysters, Gilsing, & van den Oord, 2007). As relationships deepen, trust overcomes opportunism concerns and firms are able to share more fine-grained information and tacit knowledge (Davis & Eisenhardt, 2011). In a trustful environment, alliance partners are more willing to share their proprietary knowledge without worrying about their partners taking advantages of their openness. Trustful relationships create openness within alliance portfolios, enhancing innovation appropriability. Failure to demonstrate reciprocity in relationships can lead to negative behavioral consequences such as distrust, which reduces a firm's willingness to share new ideas and knowledge (Perks & Jeffery, 2006) and become a hindrance in achieving more novel innovation (Nieto & Santamaria, 2007).

Moreover, trust allows alliance partners to divide up work without worrying about opportunistic behavior (Cowan et al., 2007). The division of work allows optimization of resources and capabilities allocation, and further foster innovation. Yet, alliance portfolios are prone to various opportunistic behaviors from alliance partners (Hanna & Walsh, 2002). The fear of unfair commercialization of new ideas often leads firms to choose suboptimal division of labor within their alliance portfolio. For partners without trust, instead of separating tasks according to competences, firms may conduct simple tasks jointly in order to monitor each other, which will decrease their efficiency significantly.

Proposition 1a: *The more alliances based on trust a firm has relative to the total number of alliances within a firm's alliance portfolio, the greater the focal firm's ability to innovate.*

Trust is not the only driver behind innovation in an alliance portfolio. Opportunism also plays a role in bringing more diversity, complementarity and risk. Opportunism can encourage firms to be more risk taking and potentially be more innovative. The development of trust and the reduction of opportunism jointly determine the successful resource sharing (Ahuja, 2000b), appropriability (Dhanaraj & Parkhe, 2006) and knowledge sharing (Nooteboom et al., 2007). Alliances built solely on trust are vulnerable to interpersonal conflicts and create issues of coordination and limit constructive thinking (Bouncken, 2011). Since trust is highly relational, social learning activities bounded with trust also follow similar behavioral codes (Bougrain & Haudeville, 2002). Firms need to find a balance between trust and opportunism in their alliance portfolio specific to their own condition and constraints. The complexity between trust and opportunism is manifested in a dilemma where partners with higher trust are more likely to exchange rich and specific information, engaging in complex problem solving, but at the same time the contribution of the close partnership rises to a threshold where the benefit of having broad market information outweighs the benefit of trust (Uzzi, 1997).

Going beyond the accepted norm that trust is always a virtue, recent studies explored the liabilities of trust by unveiling dilemmas inside trust. Molina-Morales and Martinez-Fernandez (2009) found an inverted U-shape relationship between interorganizational trust and organizations' innovation, illustrating that a medium level of trust leads to the highest level of innovation. Bidault and Castello (2009) proved that an optimum amount of mutual trust exists between alliance partners, where either too little or too much trust will hurt the joint creativity. Patzelt and Shepherd (2008) demonstrated higher trust is associated with the likelihood of persisting with underperforming alliances. Thorgren and Wincent (2011) discussed the dilemma of R&D alliances between not obtaining the desired exchange benefits without high levels of trust and suffering from negative effects and obstacles by high trust. Similarly, network studies on the role of redundant ties in building up trust and constraining opportunism, reached consensus that a redundant structure based on trust focuses more on deepening current core competences. For rich and more specific information and to broaden knowledge into new areas, trust is somewhat less important (e.g., Gulati, 1995).

> **Proposition 1b:** *The relationship between trust and innovative performance is curvilinear (inverted U-shape). At first, having alliances based on trust will impact positively a firm's innovative performance. After a certain threshold, adding more alliances based on trust to a firm's portfolio will decrease its innovative performance.*

The combination of trust and opportunism mechanisms in alliance portfolio management brings important benefits to innovation. It connects the governance perspective with the competency perspective to address the specific relational risk (Gilsing & Duysters, 2008) by solely being dependent on trust. It allows a firm to establish alliances formed upon different or even divergent motives in an attempt to reach specific innovational goals. It also grants flexibility for alliance managers to set up alliance duration with the aim of maximizing overall innovation instead of ensuring higher trust. From the governance perspective, it allows firms to adopt both formal and informal tools in managing alliance portfolios, which can be conducive to both codified and tacit knowledge cultivation and knowledge transfer. Duysters and de Man (2003) substantiated the usefulness of transitory alliances for innovation using aligned objectives other than trust as commitment under high uncertainty and environmental turbulence. Specifically, in technology-based collaboration, trust is calculative: in order to pursue higher innovation performance, firms need to optimize the benefit of trust and be wary of the downside of trust from locking into underperforming alliances (Vanhaverbeke, Gilsing, & Duysters, 2012). Based on the above evidence and arguments, it is understandable that firms will calculate opportunities and risks to determine how much trust they would like to put onto their partners and how much flexible space they want to leave for opportunism. In other words, firms adopt dual mechanisms, maintaining trust to a limited scope, pursuing their goals opportunistically and preventing their partners to take advantages opportunistically at the same time.

It is especially beneficial to embrace a temporal perspective (Das, 2006) in better understanding the relationship between trust-opportunism tension and innovation. As alliance partners continue to collaborate, trust tends to emerge and to further strengthen the collaboration (Mesquita & Lazzarini, 2008). On the one hand, terminating alliances prematurely in order to keep innovation results proprietary will underutilize the full potential of the collaboration. On the other hand, repeated interaction in terms of long alliance duration builds trust, but introduces inertia in a firm's alliance portfolio at the same time. Gilsing and Duysters (2008) contended that long-term alliance duration is undesirable, since it inhibits fast architectural innovation and kills learning potential from outside. Although having high trust is always desirable intuitively, it sets rigidity to the dynamics for alliance partner interaction. As time passes, firms get less novelty from their partners and high trust will lock them in stable but suboptimal relationships instead of seeking for novel alternatives. Therefore, long-term alliance relationships fostered by high trust are not helpful to generate innovational results.

More specifically, factors at multiple levels moderate the tension between trust and opportunism simultaneously. For example, at the macro-institutional level, Mesquita and Lazzarini (2008) pointed out that strong courts help to curb opportunistic behaviors. At the alliance portfolio level,

Wincent and his colleagues (2009) suggested that larger networks are more difficult to manage and more sensitive to opportunism risk. Kalaignanam and his colleagues (2007) mentioned that broad-scope alliances are susceptible to opportunism since they are intrinsically complex and uncertain. At the firm level, numerous studies have explored the benefits of equity alliances versus nonequity alliances in dealing with opportunism risks, proposing that the organizational embeddedness for equity alliance reduces constraints for knowledge sharing and thus contributes to novel knowledge creation and innovation (Oxley & Wada, 2009). Given the above complex multilevel moderations, the choice of optimal mix of trust and opportunism is very firm specific. If a firm can successfully leverage the tension between trust and opportunism in its alliance portfolio, it may get higher chances to perform better in innovation.

Proposition 1c: *Firms maintaining alliance portfolios characterized by a mix of alliances based on trust and opportunism will outperform firms that form alliances only based on trust or opportunism.*

STRONG TIES VERSUS WEAK TIES

One of the central tenets of literature dealing with interorganizational relationships is that the content of ties matters, a concept also known as relational embeddedness. The debate between strong and weak ties has evidenced that both of them can be beneficial for a firm's performance, although in different ways. Strong ties are associated with trust and fine-grained information exchanges between partners (Larson, 1992; Uzzi, 1997) while weak ties lead to novel information (Granovetter, 1973).

Granovetter (1973, p. 1361) defines tie strength based on a combination of the amount of time, the emotional intensity, the intimacy, and the reciprocal services which characterize the tie. In the context of tie portfolios, the focus has been typically on the frequency of interaction since the other dimensions can apply more to interpersonal rather than interorganizational relationships. For instance, the strength of a tie between two firms has been measured by the frequency of interaction between partners and their level of resource commitment to the relationship (Rowley, Behrens, & Krackhardt, 2000). Examples of strong ties include joint R&D projects and joint ventures while technology licensing falls under the weak ties category.

For the purposes of innovation performance, there is ample evidence showing that flexibility leads to success (e.g., Hargadon & Sutton, 1997). Such flexibility is provided by weak rather than strong ties. In order to be able to innovate, firms need to constantly search for new information and knowledge. According to Granovetter (1973), knowledge diffusion occurs by bridging weak ties, in other words, by weak ties that connect different social circles and provide people with access to information and resources beyond those available in their own social circle. Sociological studies have pointed to the strength of bridging weak ties. For instance, Granovetter found that among recent job changers, professional, technical, and managerial workers were more likely to hear about new jobs through weak ties (27.8%) than through strong ties (16.7%). These results suggest that weak ties generate novelty which can be regarded as a form of innovation. The above leads to the following proposition:

Proposition 2a: *The greater the number of weak ties relative to the total number of alliances within a firm's alliance portfolio, the greater the focal firm's ability to innovate.*

As previously argued, weak ties provide specific benefits since they allow firms to access novel information, an essential element for being innovative. However, it is important to consider also the quality of the information being exchanged between partners. Weak ties are more likely than strong ties to be local bridges to distant others possessing unique information. Yet, unless such ties turn into strong ones, it is difficult that the information exchanged will be of high quality. Moreover, since two partners may not trust each other, the knowledge exchanged among them may be rather explicit and thus easier to replicate. This suggests that the more bridging weak ties a firm adds to its portfolio, the greater the diversity of accessible knowledge, capabilities, and perspectives and the difficulty of integrating them (Tiwana, 2008). This tension has been characterized by Obstfeld (2005) as the problem between *idea* and *action*. In other words, to *generate* ideas, firms with bridging weak ties are better equipped toward such goal. However, for the implementation of such ideas weak ties may not be suitable due to their inherent instability. A network of collaborators with strong ties has greater capacity to *implement* innovative ideas (Tiwana, 2008). Following this logic, it may be reasonable to argue that the marginal benefits provided by adding more weak ties to a firm's portfolios will grow until a certain point after which the potential for novelty will start decreasing.

Proposition 2b: *The relationship between a focal firm's weak ties and its innovative performance is curvilinear (inverted U-shape). At first, forming weak ties will have a positive impact on a firm's innovative performance. After a certain threshold, adding more weak ties to a firm's portfolio will decrease its innovative performance.*

The impact of strong and weak ties on innovation has been usually treated as conflicting. In other words, in the search of an optimal configuration of firm's alliance portfolios, some scholars seem to suggest that weak ties while others sustain that strong ties are beneficial to innovative performance. Whereas weak ties have proved to have a greater potential for generating novel solutions (Granovetter, 1973) strong ties have also been treated as necessary for innovation and competitiveness. Put differently, while bridging weak ties provide a greater potential for innovation strong ties are better suited to implement it since they allow exchanging high-quality information and tacit knowledge. Uzzi (1996), for example, found that in the New York apparel industry firms that maintained strong ties were more able to exchange fine-grained knowledge. Teece (2000) also proposed that strong ties favor innovation. He argued that central players with strong links to suppliers and customers are better able both to drive and to benefit from innovative activities.

All in all, the above arguments seem to suggest that strong and weak ties do not compete but rather complement each other. This may be particularly true if we shift our focus from the dyad to the set of ties (i.e., the alliance portfolio). An alliance portfolio that needs to be both active in generating new ideas (i.e., having an entrepreneurial attitude) and in commercializing them may be better off if it maintains a balance between strong and weak ties. Innovative firms rely on a dynamic combination of strong and weak ties which means that they continuously reposition themselves in order to find the most suitable configuration to achieve their objectives depending on the context in which they compete (Tracey & Clark, 2003). This logic could be associated with what is known in literature as alliance ambidexterity (Gibson & Birkinshaw, 2004) which is the capacity to be simultaneously aligned and adaptive.

Proposition 2c: *Firms maintaining alliance portfolios characterized by a mix of strong and weak ties will outperform firms with portfolios characterized only by strong or weak ties.*

COMPLEMENTARITY VERSUS SIMILARITY

One of the main reasons behind alliance formation is the possibility of bringing complementary assets owned by different organizations (Nohria & Garcia-Pont, 1991) which explains why alliances have been also defined as access relationships (Stuart, 2000). Rothaermel (2001), for instance, showed that biopharmaceutical firms facilitate new product development by leveraging their partners' assets. In general terms, there is evidence of a positive relationship between a firm's number of alliances and innovation (Hagedoorn & Schakenraad, 1994; Shan et al., 1994). However, not all alliance partners are of equal value within a portfolio of alliances maintained by the focal firm. Thus, building the right relationships with the right partners becomes a key managerial challenge. For example, for a young firm or a new entrant, a partner with an established tradition in creating and launching new products successfully in a specific industry may be a more valuable partner than another partner with higher failure rates in the creation and commercialization of its products. To make sure the optimal set of partners is assembled, it needs to be investigated what a new partner can contribute to the alliance portfolio. As a consequence, this may also influence the number of ties a firm is willing to keep in its portfolio which suggests a tension between quantity and quality.

Growth in a number of firm's alliances increases potential redundancy in the alliance portfolio. Alliances are redundant to the extent they provide access to the same information (Burt, 1992) or complementary capabilities (Gomes-Casseres, 1994). Increasing the number of alliances without considering partner diversity can create inefficient configurations that return less diverse information and capabilities for greater cost than a smaller nonredundant set (Baum et al., 2000). A highly redundant configuration may also prevent access to new or novel information (Uzzi, 1996) which can limit the firm's innovativeness. Therefore, it seems logical to state that alliance portfolio configurations that provide access to more diverse information and capabilities per alliance will increase the focal firm's ability to generate novelty and launch new products. Consistent with this idea, Powell, Koput, and Smith-Doerr (1996) found that U.S. biotechnology firms with ties to a more diverse set of activities were more able to obtain resource and information-rich positions and grew more rapidly. Accordingly, we propose the following.

> **Proposition 3a:** *The greater the number of alliances with firms holding complementary assets relative to the total number of alliances within a firm's alliance portfolio, the greater the focal firm's ability to innovate.*

Following the above logic, it should be safe to argue that portfolios characterized by diverse partners generate positive innovative returns. If access to a greater diversity of information and knowledge leads to a higher rate of innovation, then why do all firms not use only diverse linkages to increase their innovative performance? This question leads naturally to a subsequent one: is there a limit to which diversity influences innovation positively? Despite the aforementioned benefits provided by alliance portfolio diversity, managing a broad set of diverse partners has some drawbacks. For instance, diversity in partner capabilities, while increasing the possible number of new recombinations, makes knowledge transfer more difficult. The greater the diversity between two firms' resource pools, the less useful is one firm's absorptive capacity for transferring knowledge and resources. Deeds and Hill (1996) illustrated that the relationship between the number of collaborative agreements and innovative performance is characterized by diminishing marginal returns. We extend these findings, by proposing that it is not just the increase in the size of the alliance portfolio but rather the complexity of managing a large set of diverse partners that will diminish the marginal returns that a firm can appropriate from its portfolio. The more diverse an alliance portfolio, the more complex is to manage it and the more management attention that is required. Furthermore, such complexity is also expected to increase the associated management costs as well as the time dedicated to the alliances. Therefore, we expect that, after a certain inflection point, the benefits of maintaining alliances with a diverse set of partners will be lower than the costs associated with managing them.

Proposition 3b: *The relationship between alliance portfolio dissimilarity and innovative performance is curvilinear (inverted U-shape). At first, having a diverse set of alliance partners will impact positively a firm's innovative performance. After a certain threshold, adding more diverse partners to a firm's portfolio will decrease its innovative performance.*

The more alliances a firm adds to its portfolio, the higher the risk of redundancy in terms of information and knowledge provided by partners compared to the firm's existing ones. This suggests that when deciding to form new alliances, firms should also look at their current alliance base. Increasing the number of alliances without considering partner diversity can create inefficient configurations that return less diverse information and capabilities for greater cost than a smaller nonredundant set (Baum et al., 2000). Entering alliances without attention to composition can also lead to conflict among a firm's partners as duplication creates rivalry among a firm's alliance partners (Gomes-Casseres, 1994).

There is a certain extent to which partnering with similar others—a phenomenon that has been also labeled as homophily—can generate positive outcomes. Several authors have argued for a positive relationship between the degree of similarity among partners in terms of assets shared in the alliance and innovation in its different forms (e.g., new knowledge and information, patents). Ahuja (2000a) found that technological similarity between alliance partners increased firm patenting after their alliance. Lane and Lubatkin (1998) argued that greater similarity in organizational properties and knowledge bases between pharmaceutical and biotech companies enhanced alliance success.

In the context of R&D alliances, partners that are too similar may experience reduced benefits from R&D collaboration. However, highly diverse partner capabilities may actually reduce the innovative benefits a firm reaps from collaborative R&D, given that firms can only assimilate capabilities that are sufficiently similar to their own (Sampson, 2007). This tension between complementarity and similarity can be resolved by embracing a paradox perspective which suggests that organizations can attend to competing demands simultaneously. Several scholars have advanced the notion that organizations can be ambidextrous (Benner & Tushman, 2003; O'Reilly & Tushman, 2004) which means they are able to manage conflicting requirements encountered. With those configurations, conflicting ingredients can coexist by adopting organizational designs that allow firms to deal with tensions. Accordingly, we propose that in the search of an optimal, and more efficient, alliance configuration—that is a configuration that provides access to more diverse information and capabilities while keeping the costs of redundancy and complexity at low levels—firms that are able to keep an equilibrium between alliances with similar partners and alliances with dissimilar alters will be able to be more innovative.

> **Proposition 3c:** *Firms maintaining alliance portfolios characterized by a mix of partners with complementary and overlapping assets will outperform firms that form alliances only with similar or complementary partners.*

DISCUSSION AND CONCLUSIONS

Adopting a paradox lens requires that we explicitly incorporate paradox theories, assumptions and methods into the alliance initiatives studied. Our goal is to make a modest contribution to the alliance portfolio literature by demonstrating how a paradox lens can allow us to conceptualize issues and problems that complement and challenge conventional contingency way of thinking. A paradox approach suggests the duality of opposition-unity by which the opposites are both conflicting and complementary at all places and at all times. This is quite different from the contingency approach which emphasizes the conditions under which certain issues are more important than others. The list of three tensions, namely (1) trust versus opportunism, (2) strong versus weak ties, and (3) complementarity versus similarity, is by no means exhaustive, but rather highlights the peculiar alliance paradoxical tension faced by firms in an attempt to stimulate further theoretical exploration along this line of research.

Empirically, we suggest to draw on multiple methodologies to test our propositions. For example "contextualized and process-oriented methodologies were often adopted to identify paradoxical tension and their management" (Smith & Lewis, 2011, p. 395). Similarly, case study is also necessary for our theory-building effort based on a relatively new phenomenon with little literature (Eisenhardt & Graebner, 2007). In particular, Smith and Lewis argued that both contradictory and interrelated decisions (i.e., paradoxical tensions) can coexist and persist over time. One way to measure these tensions, therefore, might be to trace firms' alliance history and important alliance decisions (such as partner selection, governance mode) over time.

Another issue that merits attention is how to implement solutions to overcome paradoxical tensions in order to sustain long term sustainability (Smith & Lewis, 2011). Specially, the ambidexterity literature (Benner & Tushman, 2003; Gibson & Birkinshaw, 2004; Jansen et al., 2009; Raisch & Birkinshaw, 2008), also termed as structural approach, and the temporal cycle (Puranam, Singh, & Zollo, 2006) or organizational vacillation (Boumgarden et al., 2012) perspectives, also termed as temporal approach, offer distinctive resolutions to the paradox.

The structural solutions emphasize that ambidextrous organizations can have differentiated subunits that address inconsistent demands and strategic goals (Benner & Tushman, 2003). This can be implemented through either spatial separation (Gibson & Birkinshaw, 2004) or parallel structures (Adler, Goldoftas, & Levine, 1999). For example, firms can have two separate alliance units, each responsible for creating and maintaining strong and weak ties respectively. However, this solution has been consistently criticized by its lack of dynamic element (Raish et al., 2009). For example, Raish and his colleagues (2009, p. 688) pointed out "it remains unclear how structurally differentiated units evolve over time." Given that organizations have to continuously reconfigure their activities to meet the changing environment, it is important for firms to change their structural configuration over time to achieve a temporal or dynamic fit (Montoya-Weiss, Massey, & Song, 2001; Pérez-Nordtvedt, Payne, Short, & Kedia, 2008; Shi & Prescott, 2012). It is therefore important to identify metalevel mechanisms that can provide a tool kit to direct how organizations can change differentiated structures to achieve such a dynamic fit. The identified mechanisms should provide a foundation to form process patterns that constitute a balanced state change (McGrath & Kelly, 1986) or dynamic equilibrium (Smith & Lewis, 2011) over time and do so in a systematic way.

The temporal solution, conversely, refers to temporal cycling through periods of different activities or sequentially alternating between organizational structures to deal with paradoxical goals (Raisch & Birkinshaw,

2008). This resolution includes a variety of forms such as organizational vacillation (Boumgarden et al., 2012), temporal separation (Jansen et al., 2009), sequential ambidexterity (Raisch et al., 2009), and temporal cycling (Puranam et al., 2006). For instance, firms can address the complementarity versus similarity tensions by concentrating their resources in complementary alliances within certain periods of time and shift their resource allocation to alliances with partners holding similar resources in the subsequent time periods. This solution has certainly inserted dynamic elements in the design choice (Raisch et al., 2009), but it has in the meantime downplayed the role of balance or simultaneity (Boumgarden et al., 2012). Consequently, finding synergistic integration between opposing poles is less likely (Smith & Lewis, 2011).

Our paradox approach to alliance portfolio essentially centers on the core question: "how firms and their top managers can appreciate the inherited tensions within their alliance initiatives and transfer those tensions into competitive advantage?" This approach requires managers to think and act in a paradoxical way that involves consistent inconsistency. In doing so, managers need to frequently and dynamically shift decisions and resource allocations in a creative way. Hopefully, we have stimulated some thoughts regarding how firms and their managers can incorporate a paradoxical thinking on designing and maintaining their alliance portfolio—one of the key decisions in corporate strategy.

NOTE

1. A learning race is a situation where typically one of the firms attempts to extract as much knowledge as possible from its partner while divulging as little as possible.

REFERENCES

Adler, P. S., Goldoftas, B., & Levine, D. I. (1999). Flexibility versus efficiency? A case study of model changeovers in the Toyota production system. *Organization Science*, *10*, 43-68.

Ahuja, G. (2000a). Collaboration networks, structural holes, and innovation: A longitudinal study. *Administrative Science Quarterly*, *45*, 425-455.

Ahuja, G. (2000b). The duality of collaboration: Inducements and opportunities in the formation of interfirm linkages. *Strategic Management Journal*, *21*, 317–343.

Ahuja, G., & Katila, R. (2004). Where do resources come from? The role of idiosyncratic situations. *Strategic Management Journal*, *25*, 887-907.

Andriopoulos, C., & Lewis, M. W. (2009). Exploitation-exploration tensions and organizational ambidexterity: Managing paradox. *Organization Science, 20*, 696-717.

Arrow, K. (1974). *The limits of organization*. New York, NY: Norton & Co.

Baum, J. A., Calabrese, T., & Silverman, B. S. (2000). Don't go it alone: Alliance network composition and startups' performance in Canadian biotechnology. *Strategic Management Journal, 21*, 267-294.

Benner, M. J., & Tushman, M. L. (2003). Exploitation, exploration, and process management: The productivity dilemma revisited. *Academy of Management Review, 28*, 238-256.

Bidault, F., & Castello, A. (2009). Trust and creativity: Understanding the role of trust in creativity-oriented joint developments. *R&D Management, 39*, 259-270.

Bougrain, F., & Haudeville, B. 2002. Innovation, collaboration and SMEs internal research capacities. *Research Policy, 31*, 735-747.

Boumgarden, P., Nickerson, J., & Zenger, T. R. (2012). Sailing into the wind: Exploring the relationships among ambidexterity, vacillation, and organizational performance. *Strategic Management Journal, 33*, 587-610.

Bouncken, R. B. (2011). Innovation by operating practices in project alliances—when size matters. *British Journal of Management, 22*, 586–608.

Burt, R. S. (1992). *Structural holes*. Cambridge, MA: Harvard University Press.

Cowan, R., Jonard, N., & Zimmermann, J. B. (2007). Bilateral collaboration and the emergence of innovation networks. *Management Science, 53*, 1051–1067.

Das, T. K. (2006). Strategic alliance temporalities and partner opportunism. *British Journal of Management, 17*, 1-21.

Das, T. K., & Teng, B. (2002). Alliance constellations: A social exchange perspective. *Academy of Management Review, 27*, 445-456.

Davis, J. P., & Eisenhardt, K. M. (2011). Rotating leadership and collaborative innovation: Recombination processes in symbiotic relationships. *Administrative Science Quarterly, 56*, 159-201.

De Man, A. P., & Duysters, G. (2005). Collaboration and innovation: A review of the effects of mergers, acquisitions and alliances on innovation. *Technovation, 25*, 1377-1387.

Deeds, D. L., & Hill, C. W. (1996). Strategic alliances and the rate of new product development: An empirical study of entrepreneurial biotechnology firms. *Journal of Business Venturing, 11*, 41-55.

Dhanaraj, C., & Parkhe, A. (2006). Orchestrating innovation networks. *Academy of Management Review, 31*, 659-669.

Doz, Y. L., & Hamel, G. (1998). *Alliance advantage: The art of creating value through partnering*. Boston, MA: Harvard Business School Press.

Duysters, G., & de Man, A. P. (2003). Transitory alliances: An instrument for surviving turbulent industries? *R&D Management, 33*, 49-58.

Dyer, J., Kale, P., & Singh, H. (2001). How to make strategic alliances work. *Sloan Management Review, 42*, 37-43.

Eisenhardt, K. M., & Graebner, M. E. (2007). Theory building from cases: Opportunities and challenges. *Academy of Management Journal, 50*, 25-32.

Fleming, L. (2001). Recombinant uncertainty in technological search. *Management Science, 47*, 117-132.

Gibson, C. B., & Birkinshaw, J. (2004). The antecedents, consequences, and mediating role of organizational ambidexterity. *Academy of Management Journal, 47*, 209-226.

Gilsing, V. A., & Duysters, G. M. (2008). Understanding novelty creation in exploratory networks—structural and relational embeddedness jointly considered. *Technovation, 28*, 812-817.

Goerzen, A., & Beamish, P. W. (2003). Geographic scope and multinational enterprise performance. *Strategic Management Journal, 24*, 1289-1306.

Gomes-Casseres, B. (1994). Group versus group: How alliance networks compete. *Harvard Business Review, 72*(4), 62-76.

Granovetter, M. S. (1973). The strength of weak ties. *American Journal of Sociology, 78*, 1360-1380.

Gulati, R. (1995). Does familiarity breed trust? The implications of repeated ties for contractual choice in alliances. *Academy of Management Journal, 38*, 85-112.

Gulati, R. (1999). Network location and learning: The influence of network resources and firm capabilities on alliance formation. *Strategic Management Journal, 20*, 397-420.

Gulati, R., & Higgins, M. C. (2003). Which ties matter when? The contingent effects of interorganizational partnerships on IPO success. *Strategic Management Journal, 24*, 127-144.

Hagedoorn, J. (1993). Understanding the rationale of strategic technology partnering: Interorganizational modes of cooperation and sectoral differences. *Strategic Management Journal, 14*, 371-385.

Hagedoorn, J. (2002). Inter-firm R&D partnerships: An overview of major trends and patterns since 1960. *Research Policy, 31*, 477-492.

Hagedoorn, J., & Schakenraad, J. (1994). The effect of strategic technology alliances on company performance. *Strategic Management Journal, 15*, 291-309.

Hamel, G. (1991). Competition for competence and interpartner learning within international strategic alliances. *Strategic Management Journal, 12*(S1), 83-103.

Hanna, V., & Walsh, K. (2002). Small firm networks: A successful approach to innovation? *R&D Management, 32*, 201-207.

Hargadon, A., & Sutton, R. I. (1997). Technology brokering and innovation in a product development firm. *Administrative Science Quarterly, 42*, 716-749.

Hauser, J., Tellis, G. J., & Griffin, A. (2006). Research on innovation: A review and agenda for marketing science. *Marketing Science, 25*, 687-717.

Huston, L., & Sakkab, N. (2006). Connect and develop. *Harvard Business Review, 84*(3), 58-66.

Jansen, J. J. P., Tempelaar, M., Van den Bosch, F. A. J., & Volberda, H. (2009). Structural differentiation and ambidexterity: The mediating role of integration mechanisms. *Organization Science. 20*, 797-811.

Kalaignanam, K., Shankar, V., & Varadarajan, R. (2007). Asymmetric new product development alliances: Win-win or win-lose partnerships? *Management Science, 53*, 357-374.

Kale, P., & Singh, H. (2009). Managing strategic alliances: What do we know now, and where do we go from here? *Academy of Management Perspectives, 23*(3), 45-62.

Katila, R., & Ahuja, G. (2002). Something old, something new: A longitudinal study of search behavior and new product introduction. *Academy of Management Journal, 45*, 1183-1194.

Lane, P. J., & Lubatkin, M. (1998). Relative absorptive capacity and interorganizational learning. *Strategic Management Journal, 19*, 461-477.

Larson, A. (1992). Network dyads in entrepreneurial settings: A study of the governance of exchange relationships. *Administrative Science Quarterly, 37*, 76-104.

Larsson, R., Bengtsson, L., Henriksson, K., & Sparks, J. (1998). The interorganizational learning dilemma: Collective knowledge development in strategic alliances. *Organization Science, 9*, 285-305.

Lawrence, P., & Lorsch, J. (1967). *Organizations and environment: Managing differentiation and integration*. Homewood, IL: Irwin.

Leonard-Barton, D. (1995). *Wellsprings of knowledge: Building & sustaining the sources of innovation*. Boston, MA: Harvard Business School Press.

McGrath, J. E., & Kelly, J. R. (1986). *Time and human interaction: Toward a social psychology of time*. New York, NY: Guilford Press.

Mesquita, L. F., & Lazzarini, S. G. (2008). Horizontal and vertical relationships in developing economies: Implications for SMEs' access to global markets. *Academy of Management Journal, 51*, 359-380.

Molina-Morales, F. X., & Martinez-Fernandez, M. T. (2009). Too much love in the neighborhood can hurt: How an excess of intensity and trust in relationships may produce negative effects on firms. *Strategic Management Journal, 30*, 1013-1023.

Montoya-Weiss, M. M., Massey, A. P., & Song, M. (2001). Getting it together: Temporal coordination and conflict management in global virtual teams. *Academy of Management Journal, 44*, 1251-1262.

Nielsen, B. B., & Nielsen, S. (2009). Learning and innovation in international strategic alliances: An empirical test of the role of trust and tacitness. *Journal of Management Studies, 46*, 1031-1056.

Nieto, M. J., & Santamaria, L. (2007). The importance of diverse collaborative networks for the novelty of product innovation. *Technovation, 27*, 367-377.

Nohria, N., & Garcia-Pont, C. (1991). Global strategic alliances and industry structure. *Strategic Management Journal, 12*, 105-124.

Nooteboom, B., Vanhaverbeke, W., Duysters, G., Gilsing, V., & van den Oord, A. (2007). Optimal cognitive distance and absorptive capacity. *Research Policy, 36*, 1016-1034.

O'Reilly, C. A., & Tushman, M. L. (2004). The ambidextrous organization. *Harvard Business Review, 82*(4), 74-83.

Obstfeld, D. (2005). Social networks, the tertius iungens orientation, and involvement in innovation. *Administrative Science Quarterly, 50*, 100-130.

Ostrom, E., Walker, J., & Gardner, R. (1992). Covenants with and without a sword: Self-governance is possible. *American Political Science Review, 86*, 404-417.

Oxley, J., & Wada, T. (2009). Alliance structure and the scope of knowledge transfer: Evidence from US-Japan agreements. *Management Science, 55*, 635-649.

Ozcan, P., & Eisenhardt, K. M. (2009). Origin of alliance portfolios: Entrepreneurs, network strategies, and firm performance. *Academy of Management Journal, 52*, 246-279.

Patzelt, H., & Shepherd, D. A. (2008). The decision to persist with underperforming alliances: The role of trust and control. *Journal of Management Studies, 45*, 1217-1243.

Pérez-Nordtvedt, L., Payne, T. G., Short, J. C., & Kedia, B. L. (2008). An entrainment-based model of temporal organizational fit, misfit, and performance. *Organization Science, 19*, 785-801.

Perks, H., Jeffery, R., (2006). Global network configuration for innovation: A study of international fibre innovation. *R&D Management, 36*, 67-83.

Powell, W. W., Koput, K. W., & Smith-Doerr, L. (1996). Interorganizational collaboration and the locus of innovation: Networks of learning in biotechnology. *Administrative Science Quarterly, 41*, 116-145.

Puranam, P., Singh, H., & Zollo, M. (2006). Organizing for innovation: Managing the coordination-autonomy dilemma in technology acquisitions. *Academy of Management Journal, 49*, 263-280.

Quintana-Garcia, C., & Benavides-Velasco, C. A. (2004). Cooperation, competition, and innovative capability: A panel data of European dedicated biotechnology firms, *Technovation, 24*, 927-938.

Raisch, S., & Birkinshaw, J. (2008). Organizational ambidexterity: Antecedents, outcomes, and moderators. *Journal of Management, 34*, 375-409.

Raisch, S., Birkinshaw, J., Probst, G., & Tushman, M. L. (2009). Organizational ambidexterity: Balancing exploitation and exploration for sustainable performance. *Organization Science, 20*, 685-695.

Rosen, S. M. (1994). *Science, paradox, and the Moebius principle: The evolution of a "transcultural" approach to wholeness*. New York, NY: SUNY Press.

Rothaermel, F. T. (2001). Incumbent's advantage through exploiting complementary assets via interfirm cooperation. *Strategic Management Journal, 22*, 687-699.

Rowley, T., Behrens, D., & Krackhardt, D. (2000). Redundant governance structures: An analysis of structural and relational embeddedness in the steel and semiconductor industries. *Strategic Management Journal, 21*, 369-386.

Sampson, R. C. (2007). R&D alliances and firm performance: The impact of technological diversity and alliance organization on innovation. *Academy of Management Journal, 50*, 364-386.

Schilling, M., & Phelps, C. (2007). Interfirm collaboration networks and knowledge creation: The impact of large scale network structure on firm innovation. *Management Science, 53*, 1113-1126.

Schumpeter, J. (1942). *Capitalism, socialism, and democracy*. London, UK: Unwin University Books Press.

Segrestin, B. (2005). Partnering to explore: The Renault-Nissan alliance as a forerunner of new cooperative patterns. *Research Policy, 34*, 657-672.

Shan, W., Walker, G., & Kogut, B. (1994). Interfirm cooperation and startup innovation in the biotechnology industry. *Strategic Management Journal, 15*, 387-394.

Shi, W., & Prescott, J. E. (2012). Rhythm and entrainment of acquisition and alliance initiatives and firm performance: A temporal perspective. *Organization Studies, 33*, 1281-1310.

Slaatte, H.A. (1968). *The pertinence of paradox: The dialectics of reason-in existence*. New York, NY: Humanities Press.

Smith, W. K., & Lewis, M. W. (2011). Toward a theory of paradox: A dynamic equilibrium model of organizing. *Academy of Management Review, 36*, 381-403.

Soh, P. H. (2003). The role of networking alliances in information acquisition and its implications for new product performance. *Journal of Business Venturing, 18*, 727-744.

Soh, P. H., Mahmood, I. P., & Mitchell, W. (2004). Dynamic inducements in R&D investment: Market signals and network locations. *Academy of Management Journal, 47*, 907-917.

Stuart, T. E. (2000). Interorganizational alliances and the performance of firms: A study of growth and innovation rates in a high-technology industry. *Strategic Management Journal, 21*, 791-811.

Stuart, T. E., Hoang, H., & Hybels, R. C. (1999). Interorganizational endorsements and the performance of entrepreneurial ventures. *Administrative Science Quarterly, 44*, 315-349.

Teece D. J. (2000). *Managing intellectual capital: Organizational, strategic, and policy dimensions*. Oxford, UK: Oxford University Press.

Thorgren, S., & Wincent, J. (2011). Interorganizational trust: Origins, dysfunctions and regulation of rigidities. *British Journal of Management, 22*, 21-41.

Tiwana, A. (2008). Do bridging ties complement strong ties? An empirical examination of alliance ambidexterity. *Strategic Management Journal, 29*, 251-272.

Tracey, P., & Clark, G. L. (2003). Alliances, networks and competitive strategy: Rethinking clusters of innovation. *Growth and Change, 34*(1), 1-16.

Tsai, W., & Ghoshal, S. (1998). Social capital and value creation: The role of intrafirm networks. *Academy of Management Journal, 41*, 464-476.

Uzzi, B. (1996). The sources and consequences of embeddedness for the economic performance of organizations: The network effect. *American Sociological Review, 61*, 674-698.

Uzzi, B. (1997). Social structure and competition in interfirm networks: The paradox of embeddedness. *Administrative Science Quarterly, 42*, 35-67.

Vanhaverbeke, W., Gilsing, V. A., & Duysters, G. M. (2012). Competence and governance in strategic collaboration: The differential effect of network structure on the creation of core and non-core technology. *Journal of Product Innovation Management, 29*, 784-802.

Wincent, J., Anokhin, S., & Boter, H. (2009). Network board continuity and effectiveness of open innovation in Swedish strategic small-firm networks. *R&D Management, 39*, 55-67.

CHAPTER 9

CO-INNOVATION VIA STRATEGIC ALLIANCES

The Transaction Value Perspective

Jie Zhou, Peter Ping Li, and Jun Huang

ABSTRACT

In the era of rapid globalization and innovation, strategic alliance is becoming increasingly relevant as the most critical mechanism for meeting dynamic customer demands via rapid innovation. Facing a dynamic global competition, firms have to rely on diverse and flexible alliances for the challenging goal of rapid innovation, thus making co-innovation via strategic alliances a critical mode of innovation. The extant research on co-innovation via strategic alliances is rooted in transaction cost economics (TCE) and resource-based view (RBV), while the latter also includes knowledge-based view (KBV) and dynamic capability view (DCV). Both TCE and RBV (also KBV and DCV) focus on internal hierarchy, either to minimize transaction cost (TC) or maximize hierarchy value (HV), thus neglecting hierarchy cost (HC) and transaction value (TV). Further, most studies on strategic alliances have not dealt with the balance between interfirm and intrafirm innovations. Accordingly, this chapter analyzes co-innovation via strategic alliances with transaction value perspective (TVP) so as to build an integrative framework

Strategic Alliances for Innovation and R&D, pp. 209–230

of co-innovation via strategic alliances based on TVP, which covers both intrafirm innovation (primarily for incremental innovation) and interfirm one (primarily for radical innovation); both trust and control (primarily for innovation coordination); both core focus and diverse option (primarily for innovation configuration), and finally both cost reduction and value creation as the outcome or goal of innovation.

INTRODUCTION

Global competition has reduced the role of the traditional sources of competitive advantage (Blomqvist & Levy, 2006). Increasingly, competition among firms turns on whether they can create and commercialize knowledge in a timely and cost-efficient manner (Sampson, 2007). Under the conditions of growing global competition, rapid technological advances, and frequent customer preference shifts, innovation is becoming central to competitive advantages for firms (Chandy & Tellis, 1998; Govindarajan & Kopalle, 2006). Innovation often results from the recombination of existing knowledge, but innovation also sometimes requires the creation of novel knowledge to be combined with the existing knowledge. However, in both cases of knowledge recombination and knowledge creation, a single firm often does not have the required knowledge for either recombination or creation. Few firms possess all of the required knowledge and capabilities to single-handedly develop new products (Snow & Culpan, 2012). In such a situation, *co-innovation* with other firms, which is also sometimes referred to as interfirm open innovation (Chesbrough, 2005), could be the best option available. Through cooperation with other firms, a firm can use the knowledge outside of its primary domain of expertise. Hence, interfirm cooperation between strategic alliance partners can be highly effective to acquire tacit knowledge outside its organizational boundaries. Further, firms can also enhance the speed of innovation via the intensive interfirm cooperation with strategic alliance partners.

Researchers have often characterized alliances as wellsprings of innovation (Hamel, 1991; Leonard-Barton, 1995). Strategic alliances are increasingly viewed as the most effective innovation strategy in the future. Strategic alliances have become common in many industries, especially in technology-intensive industries, including various types of alliances with customers, suppliers and even competitors (Mowery, Oxley, & Silverman, 1996). Strategic alliances can also occur between firms from totally unrelated industries (Yoshino & Rangan, 1995). Through building alliances, firms can gain access to the existing capabilities and knowledge of other firms with or without the ownership of such capabilities or knowledge (Cohen & Levinthal, 1990; Grant & Baden-Fuller, 2004). Knowledge access without ownership can apply the diverse knowledge from different firms for the development of new products or services (Buckley & Carter, 2004; Inkpen & Beamish, 1997). Knowledge access with ownership can absorb new knowledge from alliance partners as the learning opportunities and help knowledge-receiving firms to create new products or services. However, interfirm cooperation also involves risk. Alliance partners may engage in opportunistic behaviors so that they cannot be trusted, which will make cooperation difficult and sometimes impossible. The balance between intra-alliance competition and collaboration is so delicate that it requires to be managed constantly (Brandenburger & Nalebuff, 1996). Hence, we must answer a basic question with both theoretical and practical implications: *How can strategic alliance partners achieve the shared goal of co-innovation?*

In this chapter, we adopt the emerging perspective of transaction value to address the above question. First, we introduce the background of co-innovation between strategic alliance partners. Second, we review the theories to explain co-innovation, including transaction cost economics (TCE), resource-based view (RBV), and the latter's derived versions of knowledge-based view (KBV) and dynamic capability view (DCV). Further, we introduce the emerging transaction value perspective (TVP) and compare it with the other theories as to explain why TVP is uniquely positioned to best explain co-innovation between strategic alliance partners. Third, we propose an integrative framework to explain alliances co-innovation, including the two core building blocks of co-specialization and mutual trust as well as their associated antecedents and consequences. Finally, we discuss the implications of our framework for future research and practice.

BACKGROUND OF CO-INNOVATION VIA STRATEGIC ALLIANCES

Innovation is no longer just one of the sources of sustainable competitive advantage; it has become the central source. If a firm wants to survive in the long run, the most fundamental factor is innovation. In response to the rapid pace of technological advances, customer demand shifts, shortened product life cycles, and dynamic competitive advantages, firms tend to rely more on strategic alliance partners for rapid and cost-effective development of novel products and/or services by sharing and transferring complementary knowledge. In other words, innovation has become an increasingly complex process, which is increasingly involved with interfirm cooperation with external partners beyond the scope of a single firm.

Accelerating Technological Advances and Co-Innovation

Today companies face a new competitive environment. Existing technologies tend to become obsolete much more rapidly than before. Faster technological advances (Bayus, 1994) and faster implementation of existing knowledge (Borg, 2001) are adding more complexity and cost of internal innovation. Many firms can no longer afford the cost of internal R&D, so they cannot afford to own all knowledge and technologies inside a single firm. Hence, they are more inclined to share the cost of knowledge development with partners. As a result, they prefer to build longer-term cooperative relationships with other firms, such as suppliers, customers, competitors and noncompetitors from unrelated industries.

Shortened Product Life Cycles and Co-Innovation

Because of the accelerating technological advances, product life cycles have been greatly shortened. If one compares such consumer electronic products as cell-phones, digital cameras, and television, you will find the life of such products is much shorter now than before. Hence, all firms need to increase the speed of their innovation. Consequently, the competitive advantages based on existing products are fragile and vulnerable without continuous innovation (Lee, Olson, & Trimi, 2012). Further, the increasing cost of R&D cannot be easily recouped given the shortened product life cycle, which forces firms to rely more on interfirm alliances as the key source of innovation with not only enhanced knowledge base but also reduced cost burden.

Constantly Shifting Customer Needs and Co-Innovation

The traditional values to customers have been price and quality of products in the past, but customers search for novel values beyond those traditional ones. Nowadays customers pay more attention to the rich experience of value-added features, such as safety, convenience, fun, and even green or environmentally friendly and socially desirable, among others (Lee, Olson, & Trimi, 2012). Further, to develop sticky brand to maintain customer loyalty will require effective innovation in terms of both high value and low cost (Chesbrough, 2005; Lee, Olson, & Trimi, 2012). Again, it is an almost impossible task for a single firm to meet those constantly shifting customer needs. To accomplish the task of meeting the customer needs, interfirm co-innovation is becoming increasingly imperative.

Dynamic Competitive Advantage and Co-Innovation

In today's competitive environment, it is becoming difficult to sustain competitive advantages on the basis of traditional economies of scale and scope (Stel, 2011). If firms want to keep their competitive advantage, they need to renew their knowledge and advance their innovation at both faster speed and more affordable cost. Innovation is not just another option for developing competitive advantage; it has become a necessity. For fast and cost-effective innovation, firms need to greatly enhance their innovation capabilities, and there is no more effective strategy to achieve that than co-innovation via strategic alliances.

THEORIES RELATED TO CO-INNOVATION VIA STRATEGIC ALLIANCES

Alliance generally refers to all forms of interfirm cooperative mechanisms between two or more independent firms, where such firms share objectives, risks, returns and control as well as certain operational integration and mutual dependence (Bamford & Ernst, 2005). Specifically, alliances can range from weak interfirm linkages (e.g., distribution agreement or logistical supply chain relationship) to strong partnerships (e.g., equity joint ventures) (see Li, 2010, for a review). Concerning innovation, alliances have the potential for innovation by providing access to a diverse variety of capabilities and expertise (Stuart, 1998). Through alliances, firms can gain several advantages related to innovation, including direct access to complementary innovative competences, innovation risk sharing, innovation cost sharing, increased innovation flexibility, and improvement of innovative culture (Li, 2010; Stel, 2011). To understand co-innovation via strategic alliances, we need to develop a conceptual framework. To build such a framework, we will briefly review the relevant theoretical perspectives, including transaction cost economics (TCE), resource-based view (RBV), knowledge-based view (KBV), dynamic capability view (DCV), and transaction value perspective (TVP).

Transaction Cost Economics

According to TCE (Coase, 1937; Madhok, 2002; Williamson, 1993), alliances are perceived as single transactions that are pursued to overcome market failure and industrial constraints, thus primarily concerned with cost reduction rather than value creation via innovation. Hence, the central problem with TCE lies in its failure to explain the mesolevel ability of a firm to create unique values toward competitive advantages by bundling complementary resources, as argued by RBV, especially the role of knowledge as argued by KBV and the role of learning as argued by DCV (Eisenhardt & Martin, 2000; Ghoshal & Moran, 1996; Kogut & Zander, 1996; Makadok, 2001). In short, TCE is insufficient to explain alliance and co-innovation.

Resource-Based View

RBV was initially suggested by Penrose (1959), and now it has become one of the most key theories in management research. Even though it can partially explain alliance and even co-innovation via strategic alliances (Lavie, 2006), RBV is insufficient because it primarily focuses each firm as an independent entity (Barney, 1991, 2001). According to RBV, a firm is taken as a bundle of resources, which are specific to the firm within its organizational boundaries. To gain competitive advantages at the firm level, a firm should have "valuable, rare, imperfectly imitable and imperfectly substitutable" resources owned and controlled by a single firm (Maijoor & van Witteloostuijn, 1996). However, when a firm wants to innovate, it has to gain resources from other firms, especially the knowledge outside its core competence. Finally, RBV is criticized as tautological because it fails to explain where valuable resources derive from (Hoopes, Madsen, & Walker, 2003; Priem & Butler, 2001). In short, RBV is insufficient to explain alliance and co-innovation.

Knowledge-Based View

KBV partially solves the problem of tautology by explaining the role of hierarchy to create and transfer knowledge internally more effectively than the role of market, especially for nonmarketable tacit knowledge (Kogut & Zander, 1996; Nickerson & Zenger, 2004). KBV considers knowledge as the most strategically significant resource of a firm, so the primary goal of a firm is to apply the existing knowledge within the boundaries of a firm (Grant, 1996). However, KBV suffers from the same problem of focusing too much on the internal ownership of knowledge as RBV. Such a perspective is too narrow to explain alliance and co-innovation because the heterogeneous knowledge bases and capabilities distributed among diverse firms can be an unlimited source of sustained competitive advantage, and alliance is necessary for the sharing and exchange of tacit knowledge that are difficult to imitate. In other words, KBV must be expanded to include interfirm knowledge exchange and creation so as to explain co-innovation via strategic alliances. Hence, the current version of KBV has its own limitations, with the central one being the lack of power to explain why strategic alliance is better than hierarchy for competitive advantages (Hoffmann, 2007; Lavie, 2006; Li, 1998; Powell, Koput, & Smith-Doerr, 1996; Vassolo, Anand, & Folta, 2004). In short, KBV is insufficient to explain alliance and co-innovation.

Dynamic Capability View

DCV is one of the most important theories to explain innovation and learning. A firm's dynamic capabilities can be defined as the firm's ability to integrate, build and reconfigure internal and external competencies to address rapidly changing environments (Teece, Pisano, & Shuen, 1997) or as the firm's processes that use resources, specifically the processes to integrate, reconfigure, gain and release resources, to match or even to create market change (Eisenhardt & Martin, 2000). Dynamic capability has been considered as an inherent key source for a firm to gain competitive advantage in turbulent environments. Further, even firms in the market characterized by a low rate of change also need dynamic capability (Eisenhardt & Martin, 2000). Hence, DCV has even been taken as the new touchstone of performance-focused theories (Arend & Bromiley, 2009). However, DCV also suffers from its focus on the path-dependent exploitation rather than the path-breaking exploration (Schreyogg & Kliesch-Eberl, 2007). Further, DCV suffers from the same problem of focusing too much on the internal ownership of knowledge as RBV and KBV. In short, DCV is insufficient to explain alliance and co-innovation.

Transaction Value Perspective

All the above theories start with the assumptions of market failure and end up with the assumption of hierarchy benefit, so they tend to explain strategic alliance as a hybrid of market and hierarchy rather than a unique form (Li, 1998, 2010; cf. Williamson, 1991). If we want to understand co-innovation via strategic alliances, we must evoke a novel perspective. Hence, we adopt the transaction value perspective (TVP), which both challenges and supplements TCE, RBV (also KBV and DCV). While both TCE and RBV (also KBV and DCV) are different theories for internalization (Lavie, 2006), TVP is the only theory available for a new externalization distinct from the classic externalization of market transaction.

According to Zajac and Olsen (1993) who first developed the term "transaction value," there are at least two key reasons why TCE is limited in its ability to explain interorganizational strategies: (1) TCE is "essentially a single-party analysis of cost minimization," thus neglecting the interdependence between exchange partners, and (2) TCE "overemphasizes the structural analysis of interorganizational exchange relationships and neglects processual issues." In contrast, TVP offers a framework based on joint value maximization with a set of process dimensions relevant to creating value during interfirm exchange. Based on Zajac and Olsen (1993), Li (1998, 2010) expanded the notion of "transaction value" and developed it into a theory of TVP for bilateral value creation via strategic alliance to supplement TCE for unilateral cost reduction and RBV (also KBV and DCV) for unilateral value creation. TVP specifically answers the question "how is value actually created in interdependent exchange relationships" (Zajac & Olsen, 1993), especially joint knowledge creation beyond the boundary of a single firm. It is obvious that TVP complements TCE and RBV (also KBV and DCV) for a cross-level integrated theory of the firm (Li, 2010).

TVP can effectively explain co-innovation via strategic alliances because TVP considers both cost and value, and also both internal and external collaborations (Li, 1998, 2010). When we explain co-innovation via strategic alliances, we should focus on not only TC but also TV as well as both HC and HV. While TC refers to the cost related to the negative threat of external exchange due to opportunism (Williamson, 1985), TV refers to the value related to the positive opportunity of external exchange due to both co-specialization for resource configuration and mutual trust for task coordination (Li, 1998, 2010); while HC refers to the cost related to the negative effect of internal core rigidity (Leonard-Barton, 1995), HV refers to the value related to the positive effect of internal core competence (Prahalad & Hamel, 1990). While TC is related to opportunism (Williamson, 1985), HV is related to the efficiency and stability based on identity and routine (Ghoshal & Moran, 1996; Kogut & Zander, 1996). While HC is related to the imprinted inertia and bounded rationality (Verbeke, 2003), TV is related to the access to novelty and flexibility based on diverse identities and routines of diverse firms as options (Dyer & Chu, 2003; Li, 1998, 2010; Madhok, 1997; Zajac & Olsen, 1993). Hence, whether firms choose co-innovation via strategic alliances will depend on the net effect of the four factors (i.e., TC, TV, HV, and HC) in balance, with high TC and high HV pushing toward internalized innovation in contrast to high TV and high HC pushing toward co-innovation (Li, 2010).

A FRAMEWORK OF CO-INNOVATION VIA STRATEGIC ALLIANCES

Based on TVP, we posit that co-innovation via strategic alliances will involve both intrafirm and interfirm collaborations. From TVP, high TC and high HV will lead to intrafirm innovation, while high HC and high TV will lead to interfirm innovation, thus co-innovation. With the distinctive features of HV, HC, TV and TC, intrafirm innovation tends to be more effective for incremental innovation, but less effective for radical innovation as compared to interfirm co-innovation, while co-innovation tends to be more effective for radical innovation but less effective for incremental innovation than intrafirm innovation (Li, 1998, 2010). Hence, we propose a duality framework of co-innovation via strategic alliances based on TVP (see Figure 9.1).

Formal-Informal Dimension

To better understand the framework of co-innovation via strategic alliances, we must explicitly define informal elements to distinguish them from formal ones. There are few explicit definitions of formality and informality. Most scholars evoke the terms without explicitly defining them, either at the macrolevel of institutions (e.g., Peng & Heath, 1996), the mesolevel of organizations, or the microlevel of individuals (e.g., Das & Teng, 2001; Peng, Lee, & Wang, 2005; Williamson, 1993). However, the critical links between formal and informal elements are widely recognized as the core of social exchange. To differentiate formal attributes from informal ones, we adopt the definition of formality and informality proposed by Li (2005). In general, *informality* refers to the nature of social ties and events as implicitly assumed, endogenously embraced, and flexibly enforced by horizontal peer pressures in a particularistic personalized process (e.g., subjective, affective, people-oriented and sentimental), while formality as explicitly prescribed, exogenously imposed, and rigidly enforced by vertical authority powers in a universalistic depersonalized process (e.g., objective, cognitive, task-oriented and instrumental). According to Li, five basic dimensions of formality-informality can be identified: codification; formation; enforcement; power, and personalization. Judged by the above five criteria, culture, trust, social tie, relational contract and affective commitment are informal entities, while law, market, ownership and hierarchy are formal entities.

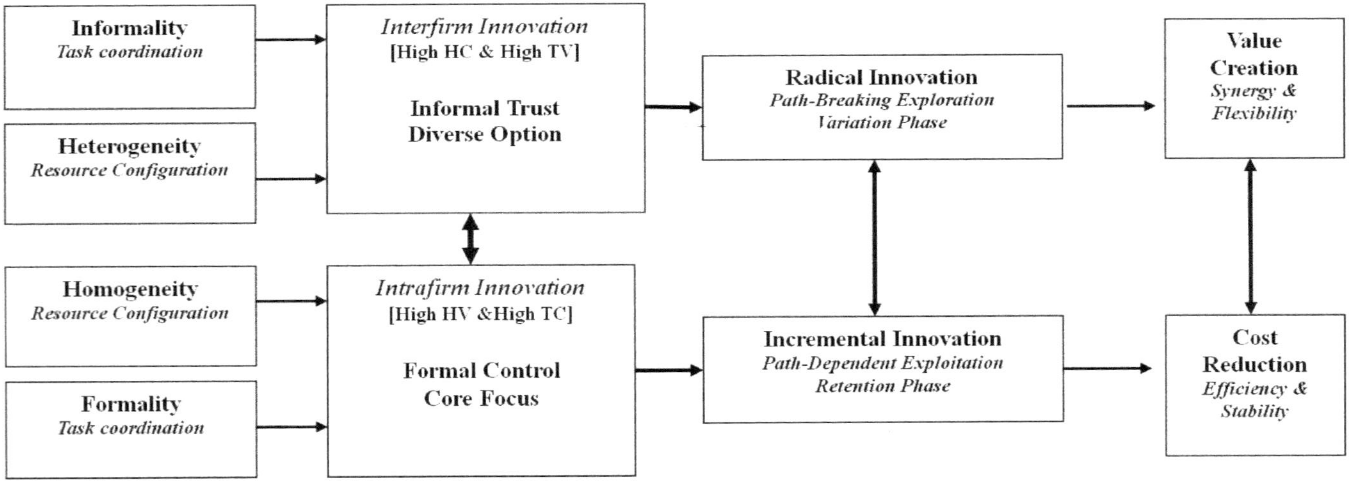

Figure 9.1. A duality framework of co-innovation via strategic alliance.

Heterogeneity-Homogeneity Dimension

When we examine co-innovation via strategic alliances, the heterogeneity-homogeneity dimension also must be delineated (Hoopes et al., 2003; Maruyama, 2004). This dimension is one of the cornerstones of RBV, including KBV and DCV (Ahuja & Katila, 2004). This dimension is closely related to intrafirm dense closure and interfirm sparse hole (Burt, 2004; Oh, Chung, & Labianca, 2004) as well as closed and open systems (Scott, 2003). Further, this dimension is central to the issue of innovation in particular, especially to open innovation (Chesbrough, 2005). *Homogeneity* refers to the composition or character of organizations or materials that are in a uniform state, while *heterogeneity* refers to the composition or character of organization or materials that are in distinctly nonuniform state (Li, 1998, 2010). For co-innovation via strategic alliances, homogeneity (as it is related to the factors of similarity, familiarity, consistency, harmony, continuity, stability, closure, and focus for efficient specialization) as well as heterogeneity (as it is related to the factors of diversity, versatility, conflict, tension, change, flexibility, openness, and novelty for effective integration) are both required. In sum, the formal-informal dimension and homogenous-heterogeneous dimension are both critical to the issue of co-innovation via strategic alliances.

Configuration-Coordination Dimension

The central issues of a firm's existence and boundary have been analyzed traditionally from the perspective of task coordination as a mode of governance in the case of TCE (Williamson, 1985), but more recently from the perspective of resource configuration as a bundle of resources in the cases of RBV (Barney, 1991, 2001), KBV (Kogut & Zander, 1996; Nickerson & Zenger, 2004), and DCV (Eisenhardt & Martin, 2000; Teece et al., 1997). However, the above two dimensions should be integrated together to explain a firm's existence and boundary, with *resource configuration* referring to the allocation scope of internal and external resources, while *task coordination* referring to the governance mode for internal and external relationships (Li, 1998 , 2010). Related to the notions of competence and governance (Nooteboom, 2004; Williamson, 1999), the core dimensions of resource configuration and task coordination are imperative for a complete theory of the firm with respect to the core issues of organizational boundary and organizational governance for both internal and external resources and relationships, thus integrating TVP with TCE and RBV (also KBV and DCV) (Carter & Hodgson, 2006; Geyskens, Steenkamp, & Kumar, 2006; Nickerson & Zenger, 2004). Specifically, the rationale for task coordination derives from cost-reduction concern, value-creation concern or both, while the rationale for resource configuration derives from focus-specialization concern, option-diversity concern or both (Li, 1998, 2010). Further, resource configuration is related to homogenous and heterogeneous resources, while task coordination is related to formal and informal tasks. Finally, the dimensions of resource configuration and task coordination delineate the three distinctive organizational forms of market, hierarchy and network (Adler, 2001; Li, 1998; Powell, 1990). In the next section, we focus on the effects of the formal-informal dimension, homogeneity-heterogeneity dimension, and configuration-coordination dimension on radical and incremental innovations as well as value creation and cost reduction via independent innovation and co-innovation.

Radical Innovation and Incremental Innovation

It is typical to differentiate radical innovation from incremental innovation as two types of innovation (Atuahene-Gima, 2005). *Incremental innovation* refers to the path-dependent type of innovation that is built upon the existing knowledge and resources, which is related to the core competence of a single firm. Hence, a single firm tends to be more effective on incremental innovation. As we mentioned before, HV and HC tend to result in and result from a path-dependent core competence or core rigidity (Danneels, 2007; Leonard-Barton, 1995; Schreyogg & Kliesch-Eberl, 2007), while TV and TC tend to result in and result from a path-breaking new core competence (Li, 1998, 2010). We argue that high HV and high TC will lead to more incremental innovation in a path-dependent process. In contrast, strategic alliance tends to be more effective on radical innovation because radical innovation requires new knowledge and resources. Hence, radical innovation tends to change a firm's old core competence, and even destroy a firm's core competence. With the distinctive mixes of HV, HC, TV and TC we mentioned before, high TV and high HC will lead to more radical innovation in a path-breaking process.

Another important issue about radical innovation and incremental innovation is how to balance these two types of innovation. Both incremental innovation and radical innovation are important for firms to sustain competitive advantages (Levinthal & March, 1993; March, 1991). However, they need different resources, different routines and different capabilities. Further, they compete for the scarce resources (Levinthal & March, 1993; March, 1991). Despite their tradeoff, the two are also complementary (Li, 1998, 2010). Hence, a firm must balance incremental and radical innovations. Specifically, the mechanisms of spatial and temporal asymmetries can help achieve such a balance. The spatial and temporal asymmetries refer to the priority a firm must give to either incremental or radical innovation at a particular given area and/or a particular given point of time. For example, when a firm is in a relatively stable context, the firm can give the priority to incremental innovation while taking radical innovation as the secondary. If a firm is in a relatively dynamic context, the firm can give the priority to radical innovation while taking incremental innovation as the secondary. Further, incremental and radical innovations require different intrafirm and interfirm resource configuration and task coordination. In general, independent innovation engaged by a single firm tends to be more effective for incremental innovation, while co-innovation via strategic alliances tends to be more effective for radical innovation.

Value Creation and Cost Reduction

The final issue in our framework is the nature of innovation in terms of value creation and cost reduction. Strategic alliance can create more radically novel products or services due to the synergy effect of pooling diverse ideas from co-specialized alliance partners with diverse core competences. Further, co-innovation can enhance the speed of innovation due to the flexibility of knowledge complementarities. In that regard, the overall effect of interfirm co-innovation can be oriented toward value creation. In contrast, the overall effect of intrafirm independent innovation can be oriented toward cost reduction.

The Configuration of Internal and External Resources

The dimension of resource configuration delineates the boundary of a focal firm to leverage its internal resources with the external resources from its network of strategic alliance partners. In contrast to a firm as a hierarchy with internal resource configuration, strategic alliance is primarily used to obtain those resources not available internally or not possible to create by a single firm alone (Das & Teng, 2000; Ireland, Hitt, & Vaidyanath, 2002; Lavie, 2006). Further, resource configuration serves as a resource bundle for two basic purposes: (1) to exploit the existing resources either available internally or leveraged externally, and (2) to explore new resources either internally or externally (Koza & Lewin, 1998; Madhok, 1997). For the purpose of exploitation, resource configuration focuses on how to benefit from the homogeneity in resource profile, cultural value and organizational routine for path-dependent learning and efficient application of existing internal and external resources (Gulati, Nohria, & Zaheer, 2000; Harrison, Hitt, Hoskisson, & Ireland, 2001). For the purpose of exploration, resource configuration focuses on how to benefit from the heterogeneity or diversity in resource profile, cultural value and organizational routine for path-breaking learning and effective application of newly-created internal and external resources (Lavie, 2006; Li, 1998; Madhok, 1997; Powell et al., 1996; Zajac & Olsen, 1993). The dimension of resource configuration can be further analyzed from the perspective of open or closed system (Scott, 2003). While closed system perspective focuses on the issues inside the boundary of a firm, open system perspective emphasizes the issues outside and spanning the boundary of a firm; while the former stresses internal homogeneity, the latter stresses external heterogeneity (Scott, 2003). Closed system stresses the necessity and inevitability of a unitary hierarchy, including specialization, identity, routine, while open system stresses the need for internal-external interaction due to the necessity of dependent embeddedness, change and diversity given the inevitable limitations of bounded rationality and imprinted inertia in a closed system.

To analyze resource configuration for co-innovation via strategic alliances, we posit that resource heterogeneity is imperative, thus implying interfirm co-specialization. When the core firm begins to build strategic alliances for innovation, they select partners according to such criteria. Alliance partners should have their own core competitive advantages, which represents co-specialization in alliance. To achieve co-innovation, co-specialization is critical because co-innovation requires more attention to radical innovation. Radical innovation needs different knowledge resources and technologies. In the co-innovation process, alliance partners should share diverse knowledge and information. Through cooperation between alliance partners, radical innovation is more likely to occur due to the greater potential of creating new knowledge through combining different ideas from different firms with different core competences. We argue co-innovation via strategic alliances as a typical form of open innovation tends to be more exploratory in nature (Chesbrough, 2005; Li, 2010), thus more likely to result in path-breaking innovations as compared to the closed innovation involving a single firm alone.

Further, we argue that resource homogeneity is central to intra-firm innovation (closed innovation). The homogeneous resources, values and management routines can enhance the exploitative efficiency of resource configuration. However, resource homogeneity also locks a firm from venturing into new business fields. Hence, internally a firm should focus on its own specialization with a core competence different from other firms. Hence, in the process of intrafirm closed innovation, firm will rely mostly on its own existing knowledge as more exploitative in nature, thus more likely to result in more incremental innovations as compared to interfirm open innovation.

Proposition 1: *Internal focus on specialization via a single firm is more critical than external option via strategic alliance for incremental innovation.*

Proposition 2: *External option via strategic alliance as open innovation is more critical than internal focus on specialization via a single firm for radical innovation.*

The Coordination of Internal and External Tasks

The dimension of task coordination delineates the governance mode adopted by a focal firm to coordinate its internal resources with the external resources from its alliance partners. In contrast to a firm as a formal hierarchy to coordinate internal resources primarily by formal control and secondarily by informal trust, a strategic alliance as an informal coalition is primarily coordinated by informal trust and secondarily by formal control (Li, 1998, 2010). Further, task coordination is designed as a mode to govern exchanges for two basic purposes: (1) to minimize the cost of exchange by focusing on unilaterally controlling the current exchange of existing resources either available internally or leveraged externally, and (2) to maximize the value of exchange by focusing on bilaterally committing to the long-term exchange of nonmarketable resources, either existing or jointly created (Das & Teng, 2001; Dyer & Chu, 2003; Li, 1998, 2010; Madhok, 1997; Zajac & Olsen, 1993). For the purpose of cost reduction, task coordination focuses on how to benefit unilaterally from exchange by protecting self-interest via formal control mechanisms, including equity ownership and detailed contract (Williamson, 1985). This is the central tenet of TCE that underlies the mainstream research on task coordination (Carter & Hodgson, 2006; Geyskens et al., 2006). For the purpose of value creation, task coordination focuses on how to benefit bilaterally by expanding the shared-interest via informal trust mechanisms, including long-term strong ties and mutual commitment (Dyer & Chu, 2003; Li, 1998, 2010; Powell et al., 1996), especially critical for R&D alliances to jointly develop new products (Doz & Hamel, 1998; Gerwin, 2004).

To analyze co-innovation via strategic alliances from the perspective of task coordination, alliance partners must be actively involved. Hence, for co-innovation, informal trust is more critical as a coordination mechanism than formal control. We posit that strong trust defined as a behavioral choice of personalized commitment to a relationship-specific interdependency between two parties to serve as the mode to govern bilateral exchange behaviors. As an anchor and an embodiment of personalized strong ties, strong trust can be an informal mode for commitment in contrast to control as a formal mode for compliance (Li, 1998, 2010). Even though alliance partners have the shared goal so as to share knowledge and information for co-innovation, some firms may take opportunistic behaviors, especially when the alliance is between competitors. When the alliance partners are competitors, some firms may have the concern of leaking out confidential information to their rivals so that they do not want to share sensitive information. In this context, strong trust is particularly critical to co-innovation.

To analyze intrafirm innovation from the perspective of task coordination, formal control is critical to any firm for its independent innovation. As a formal system, control can prevent others from engaging in negative behaviors with the threat of punishment based on the expectation of partner's self-interest seeking with guile (Williamson, 1985). Formal control tends to be more effective if it applies to the coordination within a firm as a hierarchy rather than between different firms via incomplete contracts (Williamson, 1985). Hence, intra-firm independent innovation requires more formal control than interfirm co-innovation. Certainly, trust is also necessary for intrafirm innovation, but control is more critical than trust in the case of intrafirm innovation. Similarly, control is also necessary for interfirm co-innovation, but trust is more critical than control in the case of interfirm innovation.

Proposition 3: *Internal control in a single firm is more critical than external trust in alliance partners for incremental innovation.*

Proposition 4: *External trust in alliance partners is more critical than internal control in a single firm for radical innovation.*

Co-Innovation and Independent Innovation

A firm has an internal core for efficiency and stability (Kogut & Zander, 1996; Nooteboom, 2000) because HV and HC tend to result in path-dependent core competence or rigidity (Danneels, 2007; Leonard-Barton, 1995; Schreyogg & Kliesch-Eberl, 2007). The internal core is based on organizational identity as a "central, enduring, and distinctive" set of beliefs and values "that distinguish the organization from others" (Albert & Whetten, 1985, p. 265) as well as organizational routine as the recognizable patterns of interdependent actions by multiple agents (Feldman & Pentland, 2003). These two factors render an internal collective sensemaking (Weick, 1995) so as to facilitate internal specialization and standardization for efficiency and stability. In contrast, alliance partners have inherent characteristics due to an external scope for synergy and flexibility (Hoffmann, 2007; Lavie, 2006; Nooteboom, 2000; Vassolo et al., 2004), TV and TC tend to result in a path-breaking new core competence (Li, 1998, 2010) despite the risk of opportunism (Williamson, 1985). Interfirm synergy derives from novel combinations of diverse resources from distinctive firms (Ahuja & Katila, 2004), while interfirm flexibility derives from informal combinations of compatible resources (Li, 1998, 2010). While internal resources tend to be tightly-coupled with respect to configuration and coordination due to shared identity and routine, external resources leveraged via alliances tend to be loosely-coupled due to unique novelty and informality. With the distinctive associations of HV, HC, TV and TC, we argue that intrafirm independent innovation tends to be more effective for incremental innovation while less effective for radical innovation than interfirm co-innovation; also, while interfirm co-innovation tends to be more effective for radical innovation, it is less effective for incremental innovation than intra-firm independent innovation.

> **Proposition 5:** *Intra-firm independent innovation is more effective for incremental innovation, but less effective for radical innovation.*

> **Proposition 6:** *Interfirm co-innovation is more effective for radical innovation, but less effective for incremental innovation.*

CONCLUSION

Due to the increasingly competitive environment, innovation has been more imperative for firms than before, but firms can never have enough knowledge and resources to innovate everything alone. Hence, co-innovation via strategic alliances has been a viable option for the goal of sustainable innovation. This chapter adopts TVP and applies it to co-innovation via strategic alliances. TVP complements TCE and RBV (also KBV and DCV) by adding new insights into the unique effectiveness of co-innovation via strategic alliances. In general, TVP can explain co-innovation via strategic alliances better than other extant theories. In particular, we have examined the limitations of the extant theories of TCE and RBV (also KBV and DCV) with respect to their neglect of the value-creating potential of strategic alliances. By shifting the focus from the level of a single firm to the cross-level of both firm and interfirm alliance, we have moved from an internal focus to an integrated focus on the internal-external integration. This shift of focus and level has the potential to advance our understanding about co-innovation via strategic alliances. In particular, we have proposed an integrative framework to explain co-innovation, including the core building blocks of co-specialization and mutual trust for the boundary and nature of the firm in addition to the related dimensions of formality-informality, homogeneity-heterogeneity, cost reduction and value creation, radical and incremental innovations, and intra-firm and interfirm innovations.

For future research, we posit that TVP is an important lens to analyze co-innovation via strategic alliances. In particular, the focus of analysis should shift from firm level or dyadic level to a cross-level of both firm and interfirm network because the literature has unduly neglected the interplay between firm and network of strategic alliance. This shift is made possible by the introduction of TVP. How to balance incremental innovation and radical innovation between a focal firm and strategic alliance partners as a cross-level issue can be an important research topic for future research. Further, we need to refine and test the proposed framework of co-innovation via strategic alliances, including longitudinal case studies and large-sample tests.

REFERENCES

Adler, P. S. (2001). Market, hierarchy and trust: The knowledge economy and the future of capitalism. *Organization Science, 12*, 215-234.

Ahuja, G., & Katila, R. (2004). Where do resources come from? The role of idiosyncratic situations. *Strategic Management Journal, 25*, 887-907.

Albert, S., & Whetten, D. (1985). Organizational identity. *Research in Organizational Behavior*. *14*, 263-295.

Arend, R. J., & Bromiley, P. (2009). Assessing the dynamic capabilities view: Spare change, everyone? *Strategic Organization*, *7*, 75-90.

Atuahene-Gima, K. (2005). Resolving the capability-rigidity paradox in new product innovation. *Journal of Marketing*, *69*(4), 61-83.

Bamford, J., & Ernst, D. (2005). Governing joint ventures. *McKinsey Quarterly*, Special Edition, 62-68.

Barney, J. B. (1991). Firm resources and sustained competitive advantage. *Journal of Management*, *17*, 99-120.

Barney, J. B. (2001). Is the resource-based theory a useful perspective for strategic management research? Yes. *Academy of Management Review*, *26*, 41-56.

Bayus, B. L. (1994). Are product life cycles really getting shorter? *Journal of Product Innovation Management*, *11*, 300-308.

Blomqvist, K., & Levy, J. (2006). Collaboration capability – a focal concept in knowledge creation and collaborative innovation in networks, *International Journal Management Concepts and Philosophy*, *2*(1), 31-48.

Borg, E. A. (2001). Knowledge, information and intellectual property: Implications for marketing relationships. *Technovation, 21*, 515-524

Brandenburger, A. M., & Nalebuff, B. J. (1996). *Co-opetition*. New York, NY: Doubleday-Bantam.

Buckley, P. J., & Carter, M. J. (2004). A formal analysis of knowledge combination in multinational enterprises. *Journal of International Business Studies, 35*, 371-384.

Burt, R. S. (2004). Structural holes and good ideas. *American Journal of Sociology, 110*, 349-400.

Carter, R., & Hodgson, G. M. (2006). The impact of empirical tests of transaction cost economics on the debate on the nature of the firm. *Strategic Management Journal, 27*, 462-476.

Chandy, R. K., & Tellis, G. J. (1998). Organizational for radical product innovation: The overlooked role of willingness to cannibalize. *Journal of Marketing Research, 35*, 474-487.

Chesbrough, H. W. (2005). *Open innovation: The new imperative for creating and profiting from technology*. Boston, MA: Harvard Business Press.

Coase, R. E. (1937). The nature of the firm. *Economica, 4*, 386-405.

Cohen, W. M., & Levinthal, D. A. (1990). Absorptive capacity: A new perspective on learning and innovation. *Administrative Science Quarterly, 35*, 128-152.

Danneels, E. (2007). The process of technological competence leveraging. *Strategic Management Journal, 28*, 511-533.

Das, T. K., & Teng, B. (2000). A resource-based theory of strategic alliances. *Journal of Management, 26*, 31-61.

Das. T. K., & Teng, B. (2001). Trust, control, and risk in strategic alliances: An integrated framework. *Organization Studies, 22*, 251-283.

Doz, Y. L., & Hamel, G. (1998). *Alliance advantage: The art of creating value through partnering*. Boston, MA: Harvard Business School Press.

Dyer, J. H., & Chu, W. (2003). The role of trustworthiness in reducing transaction costs and increasing information sharing: Empirical evidence from the United States, Japan, and Korea. *Organization Science. 14*, 57-68.

Eisenhardt, K. M., & Martin, J. A. (2000). Dynamic capabilities: What are they? *Strategic Management Journal, 21*, 1105-1121.

Feldman, M. S., & Pentland, B. T. (2003). Reconceptualizing routines as a source of flexibility and change. *Administrative Science Quarterly, 48*, 94-118.

Gerwin, D. (2004). Coordinating new product development in strategic alliances. *Academy of Management Review, 29*, 241-257.

Geyskens, I., Steenkamp, J. E., & Kumar, N. (2006). Make, buy, or ally: A transaction cost theory meta-analysis. *Academy of Management Journal, 49*, 519-543.

Ghoshal, S., & Moran, P. (1996). Bad for practice: A critique of the transaction cost theory. *Academy of Management Review, 21*, 13-47.

Govindarajan, V., & Kopalle, P. K. (2006). Disruptiveness of innovations: Measurement and an assessment of reliability and validity. *Strategic Management Journal, 27*, 189-199.

Grant, R. M. (1996). Toward a knowledge-based theory of the firm. *Strategic Management Journal, 17* (Winter special issue), 109-122.

Grant, R. M., & Baden-Fuller, C. (2004). A knowledge accessing theory of strategic alliances. *Journal of Management Studies, 41*, 61-84.

Gulati, R., Nohria, N., & Zaheer, A. (2000). Strategic networks. *Strategic Management Journal, 21*, 203-215.

Hamel, G. (1991). Competition for competence and interpartner learning within international strategic alliances. *Strategic Management Journal, 12*(Summer special issue), 83-103.

Harrison, J. S., Hitt, M. A., Hoskisson, R. E., & Ireland, R. D. (2001). Resource complementarity in business combinations: Extending the logic to organizational alliances. *Journal of Management*, 27, 679-690.

Hoffmann, W. H. (2007). Strategies for managing a portfolio of alliances. *Strategic Management Journal, 28*, 827-856.

Hoopes, D. G., Madsen, T. L., & Walker, G. (2003). Why is there a resource-based view? Toward a theory of competitive heterogeneity. *Strategic Management Journal, 24*, 889-902.

Inkpen, A. C., & Beamish, P. W. (1997). Knowledge, bargaining power, and the instability of international joint ventures. *Academy of Management Review, 22*, 177–202.

Ireland, R. D., Hitt, M. A., & Vaidyanath, D. (2002). Alliance management as a source of competitive advantage. *Journal. of Management, 28*, 413-446.

Kogut, B., & Zander, U. (1996). What firms do? Coordination, identity, and learning. *Organization Science, 5*, 502-518.

Koza, M. P., & Lewin, A. Y. (1998). The co-evolution of strategic alliances. *Organization Science, 9*, 255-264.

Lavie, D. (2006). The competitive advantage of interconnected firms: An extension of the resource-based view. *Academy of Management Review, 31*, 638-658.

Lee, S. M., Olson, D. L., & Trimi, S. (2012). Co-innovation: Convergenomics, collaboration, and co-creation for organizational values. *Management Decision, 50*, 817-831.

Leonard-Barton, D. (1995). *Wellsprings of knowledge: Building and sustaining the sources of innovation*. Boston, MA: Harvard Business School Press.

Levinthal, D. A., & March, J. G. (1993). The myopia of learning. *Strategic Management Journal, 14*, 95-112.

Li, P. P. (1998). Towards a geocentric framework of organizational form: A holistic, dynamic and paradoxical approach. *Organization Studies, 19*, 829-861.

Li, P. P. (2005). The puzzle of China's township-village enterprises: The paradox of local corporatism in a dual-track economic transition. *Management and Organization Review, 1*, 197-224.

Li, P. P. (2010). Toward a learning-based view of internationalization: The accelerated trajectories of cross-border learning for latecomers. *Journal of International Management, 16*, 43-59.

Madhok, A. (1997). Cost, value and foreign market entry mode: The transaction and the firm. *Strategic Management Journal, 18*, 39-61.

Madhok, A. (2002). Reassessing the fundamentals and beyond: Ronald Coase, the transaction cost and resource-based theories of the firm and the institutional structure of production. *Strategic Management Journal, 23*, 535-550.

Maijoor, S., & Van Witteloostuijn, A. (1996). An empirical test of the resource-based theory: Strategic regulation in Dutch audit industry. *Strategic Management Journal, 17*, 549-569.

Makadok, R. (2001). Toward a synthesis of the resource-based and dynamic-capacity views of rent creation. *Strategic Management Journal, 22*, 387-401.

March, J. G. (1991). Exploration and exploitation in organizational learning. *Organization Science, 2*, 71-87.

Maruyama, M. (2004). Polyocular vision or subunderstanding? *Organization Studies, 25*, 467-480.

Mowery, D. C., Oxley, J. E., & Silverman, B. S. (1996). Strategic alliances and inter firm knowledge transfer. *Strategic Management Journal, 17*, 71-91.

Nickerson, J. A., & Zenger, T. R. (2004). A knowledge-based theory of the firm—The problem-solving perspective. *Organization Science, 15*, 617-632.

Nooteboom, B. (2004). Governance and competence: How can they be combined? *Cambridge Journal of Economics, 28*, 505-526.

Nooteboom, B. (2000). *Learning and innovation in organizations and economies*. New York, NY: Oxford University Press.

Oh, H., Chung M., & Labianca, G. (2004). Group social capital and group effectiveness: The role of informal socializing ties. *Academy of Management Journal, 47*, 860-875.

Peng, M. W., & Heath, P. S. (1996). The growth of the firm in planned economies in transition: Institutions, organizations, and strategic choice. *Academy of Management Review, 21*, 492-528.

Peng, M. W., Lee, S., & Wang, D. (2005). What determines the scope of the firm over time? A focus on institutional relatedness. *Academy of Management Review, 30*, 622-633.

Penrose, E. G. (1959). *The theory of the growth of the firm*. New York, NY: Wiley.

Powell, W. W. (1990). Neither market nor hierarchy: Network forms of organization. *Research in Organizational Behaviour, 12*, 295-336.

Powell, W. W., Koput K. W., & Smith-Doerr L. (1996). Interorganizational collaboration and the locus of innovation: Networks of learning in biotechnology. *Administrate Science Quarterly, 41*, 116-145.

Prahalad, C. K., & Hamel, G. (1990). The core competence of the organizations. *Harvard Business Review, 90*(3), 79-91.

Priem, R. L., & Butler, J. E. (2001). Is the resource-based theory a useful perspective for strategic management research? *Academy of Management Review, 26*, 22-40.

Sampson, R. C. (2007). R&D alliances and firm performance: The impact of technological diversity and alliance organization on innovation. *Academy of Management Journal, 50*, 364-386.

Schreyogg, G., & Kliesch-Eberl, M. (2007). How dynamic can organizational capabilities be? Towards a dual-process model of capability dynamization. *Strategic Management Journal, 28*, 15-24

Scott, W. R. (2003). *Organizations: Rational, natural and open systems* (5th ed.) Eaglewood Cliff, NJ: Prentice-Hall.

Stel, F. (2011). *Improving the performance of co-innovation alliances: Cooperating effectively with new business partners* (Unpublished PhD dissertation). Tilburg University, Netherlands.

Snow, C. C., & Culpan, R. (2012). Open innovation through a collaborative community of firms: An emerging organization design. In T. K. Das (Ed.), *Strategic alliances for value creation* (pp. 279-300). Charlotte, NC: Information Age Publishing.

Stuart, T. (1998). Network positions and propensities to collaborate: An investigation o f strategic alliance formation in a high-technology industry. *Administrative Science Quarterly, 43*, 668-698.

Teece, D. J., Pisano G., & Shuen, A. (1997). Dynamic capabilities and strategic management. *Strategic Management Journal, 18*, 509-533.

Vassolo, R. S., Anand, J., & Folta, T. B. (2004). Non-additivity in portfolios of exploration activities: A real options-based analysis of equity alliances in biotechnology. *Strategic Management Journal, 25*, 1045-1061.

Verbeke, A. (2003). The evolutionary view of the MNE and the future of internalization theory. *Journal of International Business Studies, 34*, 498-504.

Weick, K. E. (1995). *Sensemaking in organizations*. Thousand Oaks, CA: Sage.

Williamson, O. E. (1985). *Economic institutions of capitalism*. New York, NY: Free Press.

Williamson, O. E. (1991). Strategizing, economizing, and economic organization. *Strategic Management Journal, 12*(Winter Issue), 75-94.

Williamson, O. E. (1993). The logic of economic organization. In O. E. Williamson & S. G. Winter (Eds.), *The nature of the firm: Origins, evolution, and development* (pp. 90-116). New York, NY: Oxford University Press.

Williamson, O. E. (1999). Strategy research: Governance and competence perspectives. *Strategic Management Journal, 20*, 1087-1108.

Yoshino, M. Y., & Rangan, U. S. (1995). *Strategic alliances: An entrepreneurial approach to globalization*. Boston, MA: Harvard Business School Press.

Zajac, E. J., & Olsen, C. P. (1993). From transaction cost to transaction value analysis: Implications for the study of inter-organizational strategies. *Journal of Management Studies, 30*, 131-145.

CHAPTER 10

STRATEGIC ALLIANCES FOR INNOVATION IN CHINA

Institutional Alliance Capability?

Mark Greeven and Geerten van de Kaa

ABSTRACT

Despite widespread uncertainty and opportunism, limitations in factor markets and higher search costs for critical resources, a private sector is rapidly emerging in China with new business models, new organizational forms, and the development of new products and technologies. In this chapter we explore strategic alliances for innovation in the context of China by, first, reviewing extant literature and, second, exploring two illustrative cases. Alliances of Chinese private firms are understudied, despite the significance of domestic private firms in the economy and the widespread role of a variety of alliances in mitigating institutional shortcomings. We propose that the notion of alliance capability may be usefully employed in studying Chinese alliances. Furthermore, our cases suggest that the literature on alliances may incorporate institutional alliances as an element for explaining alliance capability in China's emerging economy context.

Strategic Alliances for Innovation and R&D, pp. 231–247

INTRODUCTION

A private sector is rapidly emerging in China with new business models, new organizational forms, and the development of new products and technologies. The emergence of new industries and the commercialization of knowledge-intensive products such as TD-SCDMA, pictographic language software and new antimalaria drug suggest that China is becoming more innovative. Emerging economies such as China's are often characterized by widespread uncertainty and opportunism, limitations in factor markets and higher search costs for critical resources (Ahlstrom & Bruton, 2010; Roth & Kostova, 2003). Current literature emphasizes the role of various kinds of formal and informal alliances and business networks and acknowledges the strategic role of the central and local government in these networks for innovation in China (e.g., Perks, Khan, & Zhang, 2009; Siu & Bao, 2008; Xin & Pearce, 1996). These studies emphasize a set of informal arrangements to overcome the assumed shortcomings in the business environment (Xin & Pearce, 1996), but remain inconclusive on how these informal arrangements affect innovation. In this chapter we will draw on the strategic alliance literature and explore the notion of alliance capability (Kale & Singh, 1999) in order to extend our understanding of innovation in China and also extend the use of the alliance capability concept in a new institutional context.

The strategic alliance literature has emphasized the role of networks, alliances and interpartner learning for innovation (Gulati, 1998; Hamel, 1991; Hamel, Doz, & Prahalad, 1989). It is argued that a firm can learn from actors and obtain external knowledge, increasing its innovation output (Powell, Koput, & Smith-Doerr, 1996). Within this context, developing an alliance capability may give a competitive advantage. After reviewing the broad literature on alliance capability we identified two areas of contribution. Alliance capability in China may be different due to institutional differences, as suggested by Hitt, Ahlstrom, Dacin, Levitas, and Svobodina (2004). Furthermore, there is limited research on alliance capability in China that focuses on foreign firms (e.g., Murray, Kotabe, & Zhou, 2005). However, China's economy is driven by the private sector and private firms can no longer be ignored as significant economic actors; we therefore focus on alliance capability of Chinese private firms.

In our empirical study we identify what kind of alliances Chinese new ventures have initiated with what kind of actors and show that these new ventures have a variety of external formal and informal alliances to allow them to mobilize resources for innovation. Our empirical cases suggest that private firms develop strategic alliances with both business and government in order to mobilize resources for innovation. In conclusion, we suggest alliance capability as a helpful concept in understanding innovation in China and extend the notion of alliance capability to include an institutional element.

Our main objective is to explore the extent to which Chinese firms have formed an alliance capability. In the next section we review the literature on alliance capability in general and attempt to apply this to the case of emerging economies such as China. In the third section we describe the methods applied in this study. In the fourth section we present the outcomes of an extensive literature review on alliance and alliance capability and identify the different studies that focus on alliances in China. In the literature we find few publications that focus on interorganizational relationships *between* Chinese firms and no study deals with alliance capability in China as such. In the fifth section we present two illustrative case studies of alliances in China, identifying the role of strategic alliances with business and government partners, suggesting an institutional alliance capability. The last section contains a discussion and conclusion.

ALLIANCE CAPABILITY

An alliance capability "is defined as a higher-order resource which is difficult to obtain or imitate and has the potential to enhance the performance of the firm's alliance portfolio" (Heimeriks & Duysters, 2007, p. 30). Alliance capability refers not only to the structure and fit of alliances but shifts attention to alliance experience (Anand & Khanna, 2000) and learning (Kale, Dyer, & Singh, 2002). In fact, one of the main reasons that alliances are formed by firms lies in the fact that firms can learn know-how and capabilities from other partners in the alliance (Kale, Singh, & Perlmutter, 2000). Mutual trust and interaction between alliance members is the basis for the creation of relational capital and facilitates learning and knowledge transfer (Kale et al., 2000). Some researchers elaborate on specific practices that firms may apply to form an alliance capability (Draulans, de Man, & Volberda, 2003). These include "organizational processes facilitating the accumulation, codification and sharing of alliance know-how embedded in the firm's alliance experience" (Kale & Singh, 1999, p. 1). It is argued that "differences in sources of alliance capabilities explain performance heterogeneity" and that an alliance capability may mediate the effect of alliance experience on alliance performance (Heimeriks & Duysters, 2007, p. 25). In particular, in emerging economies, where firms may lack experience and continuous change requires strong learning capabilities, the notion of alliance capability may be powerful.

Research emanating from the resource-based view of the firm has emphasized the concept of alliance competence (Lambe, Spekman, & Hunt, 2002; Sanchez, Heene, & Thomas, 1996) which is comparable to alliance capability. Scholars who have explored the concept of alliance competence argue that it "contributes to alliance success, both directly and through the acquisition and creation of resources" (Lambe et al., 2002, p. 141). This seems to be particularly relevant for the Chinese context where companies face a resource poor and ambiguous environment.

METHODS

In order to explore alliance capability of Chinese firms, we first evaluated the extent to which the literature focusing on alliance capability has studied alliances or interorganizational relationships in the Chinese context. This may provide us with a first indication of the extent to which an alliance capability has been employed in the Chinese context. Next, in a separate study, we illustrate two cases of alliance capability in China.

We started the first study with an extensive search for papers that focus on factors for alliance capability. We started with a search using different databases including ISI web of knowledge, Scopus, and Google Scholar. We used different search terms including alliance, alliance capabilities, alliance formation, joint venture, and so forth, following Hagedoorn (1990). For every publication found we analyzed each of the sources to which that publication referred to and we analyzed the papers that cited the publication. This resulted in a long list of 144 publications.

To categorize the publications we used the following dimensions: Title, Journal, Summary (Abstract), Geographical context (United States, Europe, Japan, China, etc.), Case (IT, CE, Biotech, etc.), Type of paper (literature review/conceptual paper/case studies/experiment/survey/etc.), Type of actors involved (firms, institutions, regulatory agencies, consumers, etc.), Type of alliance 1 (alliance, consortium, acquisition, etc. (see Hagedoorn, 1990, for an overview), Type of alliance 2 (Foreign firm-foreign firm; foreign firm-Chinese firm; Chinese firm-Chinese firm), and factors affecting alliance formation (product-market diversity of firm, industry convergence, market uncertainty, etc.). Using these categorizations we arrived at three distinct datasets: alliances in China, alliances outside of China, alliances between Chinese firms and outside parties. These datasets are available upon request. We concentrate here on alliances in China.

ALLIANCES IN CHINA

Publications that focus on alliances and alliance capability have been published in a diverse range of journals. In Table 10.1, a list of the journals is presented. Not surprisingly, most publications have been published in management related journals and 52% of the publications on alliance capability that we found were published in one of four journals: *Strategic Management Journal, Academy of Management Journal, Organization Science, and Journal of Management Studies* (see Table 10.1).

Table 10.1. Dataset by Journal Publication

	Full Dataset	*Alliances in China*
Strategic Management Journal	32	4
Academy of Management Journal	16	4
Organization Science	13	3
Journal of Management Studies	13	3
Journal of International Business Studies	9	4
Research Policy	7	4
Organization Studies	7	1
Academy of Management Review	6	
Administrative Science Quarterly	5	
Journal of Management	5	1
Harvard Business Review	3	
Management Science	3	
Journal of World Business	2	2
MIT Sloan Management Review	2	
Academy of Management Perspectives	1	
Asia Pacific Journal of Management	1	1
California Management Review	1	
Industrial and Corporate Change	1	
Industrial Marketing Management	1	
Journal of Applied Psychology	1	1
Journal of Business Venturing	1	
Journal of Economic Behavior and Organization	1	
Journal of Engineering and Technology Management	1	
Journal of Industrial Economics	1	
Journal of Marketing	1	
Journal of Marketing Research	1	
Journal of Product Innovation Management	1	

(Table continues on next page)

Table 10.1. Dataset by Journal Publication (Cont.)

	Full Dataset	*Alliances in China*
Journal of the Academy of Marketing Science	1	
Managerial and Decision Economics	1	
McKinsey Quarterly	1	
R&D Management	1	
RAND Journal of Economics	1	
Technology Analysis and Strategic Management	1	
The Journal of Industrial Economics	1	
The Journal of Product Innovation Management	1	1
Total	144	29

Interestingly, 48% of the publications on alliance capability that focus on the alliances between Chinese firms were published in the same set of journals indicating that alliances in China is not a niche subject and is published in general management journals. Interestingly, Table 10.1 also shows that from the 144 publications on alliances in general, 29 focused on alliances between Chinese firms. In the remainder of this section we will concentrate on the 29 papers that focus on alliances in China.

Figure 10.1 shows the type of alliances that are studied in the papers that focus on alliances in China. As some publications focus on multiple types of interorganizational relations, we have included all of these types. What is striking from this comparison is that we find a wide variety of types of alliances. Interestingly, most interorganizational relationships fall in the category joint venture, strategic alliances or customer-supplier relations. Moreover, not one study deals explicitly with the alliance capability notion in the Chinese context. In particular, 13 (43%) of the papers deals with joint ventures. This is not entirely surprising, as most research published in international journals is dealing with Sino-foreign joint ventures and the determinants of their success and failure. Moreover, after opening up and until the 1990s foreign firms were required by the Chinese government to establish a joint venture structure to be able to enter the Chinese market.

The second dimension of alliance we investigated is origin of the partner: Chinese versus foreign. Figure 10.2 shows that 8 (27%) of the papers deal with Chinese-Chinese alliances, whereas 22 (73%) of the papers deal with foreign-Chinese alliances. It appears that research on Chinese-Chinese alliances is underdeveloped, in particular given the emphasis being given to the role of networks, guanxi and informal relationships in doing

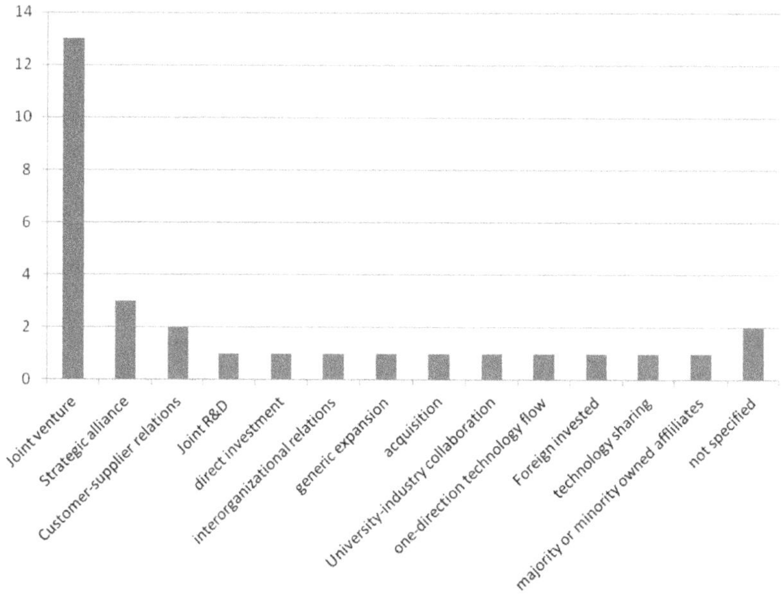

Figure 10.1. Types of alliances.

business in China. Since the mid-1990s a significant private sector has emerged in China (Tsui, Bian, & Cheng, 2006). Less the result of state initiated mass privatization, as in the case of Eastern Europe and Russia, the Chinese private business sector emerged via ownership changes in SOEs, the rise of TVEs, and other hybrid forms of ownership and via entrepreneurship when individuals founded new firms (Tsui et al., 2006). By 2011, the private sector had become not only the main generator of output (an estimated 70% of GDP) and employment, but also the most innovative sector in China (World Bank, 2012). This unprecedented growth of the private business sector enabled China to develop in a competitive market economy.

Further, we looked at the context of the alliance: what is the sector under investigation? Figure 10.3 shows that the majority of the papers deals with high-tech and manufacturing sectors, while almost half of the papers do not have a particular sector-context. Comparing these results with the full data set of papers on alliance capability outside China, we see that telecom and biotechnology are underrepresented, even as strategic alliances are widespread in these sectors. There may be several reasons for this. First, both sectors are new in China and just recently emerging. Second, the telecom sector in China is heavily regulated and dominated by a small selection of state-owned firms.

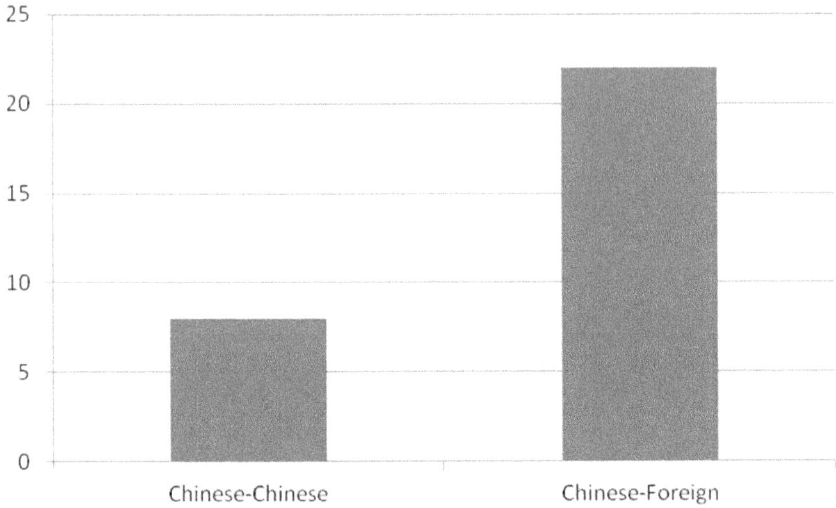

Figure 10.2. Origin of the partner: Chinese versus foreign.

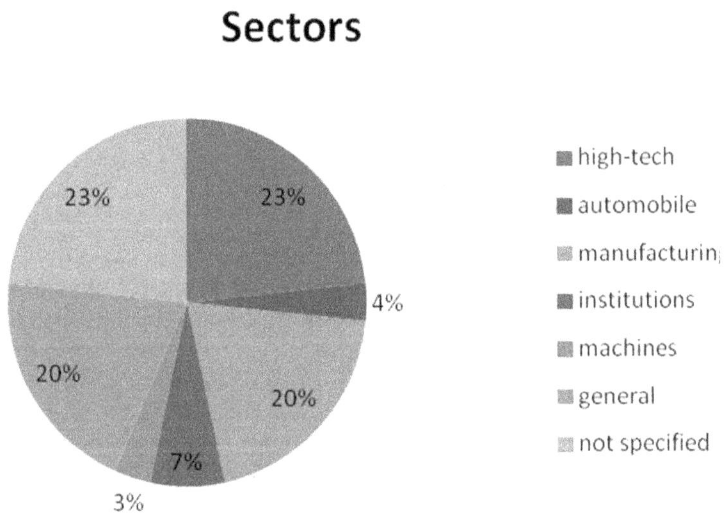

Figure 10.3. Context of the alliance.

Also, we looked at the methodologies employed in the studies on Chinese alliances (see Figure 10.4). The majority of papers employs a survey method in which existing theories and hypotheses are tested within the Chinese context. About 20% of the papers employs a case study method, while one paper is conceptual. We believe that more case studies may shed light on the variety of alliance activities by Chinese firms beyond joint ventures with foreign firms.

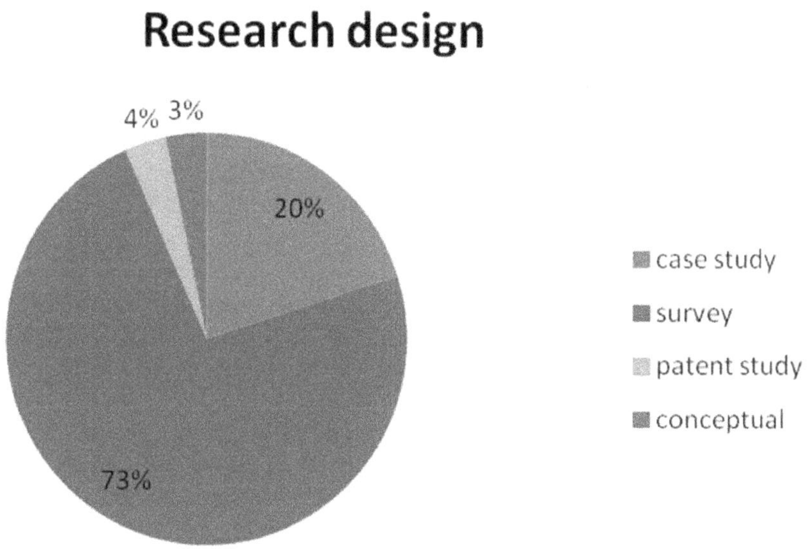

Figure 10.4. Methodologies employed in the studies on Chinese alliances.

Lastly, we looked at the factors that contributed to alliance formation in China. Various factors are mentioned in the literature. Interestingly, some studies also mention factors for alliance formation in china that are also used in the literature on alliance formation in general, including network learning and enhancing the competitive position. Indeed, scholars in the area of alliance formation argue that from a strategy perspective firms establish relationships to enhance their competitive position (Eisenhardt & Schoonhoven, 1996; Kogut, 1988; Porter & Fuller, 1986) and learn from each other (Hamel, 1991; Hamel et al., 1989).

In what follows we present two illustrative case studies of Chinese private firms who engage in alliance activities for innovation. The goal of these cases is to explore a variety of alliance activities and provide insights for future research. The names of the companies are fictive on request of the interviewees.

CASE STUDIES OF CHINESE PRIVATE FIRMS

Zhejiang HeYi Case

Established in October 2000, Zhejiang HeYi is a professional investment and management establishment specializing in the field of Internet profession and information technology. The company presently operates a number of well-known Internet sites, including China's retail industry portal, simulation of the world's first virtual city website, Chinese store trading portal, China's only real estate search engine, and Chinese martial arts and cultural portal.

WebE, the dominant information web provider in the China's retail industry, was launched in January, 2001. WebE has consistently provided reliable information to the mass retail stores through its established website. YongJin was launched in July, 2003. A few months later, in November, the website received authorization from Mr. Yong Jin to be his only official site in China. YongJin was being positioned and labeled as China's Martial Arts portal. In 2004 they acquired a local company which provided BBS and was called Hangzhou Home with the help of local government investments. At that time, this company was the biggest online community company, with 20 employees. Furthermore, the company took over Aldin providing 3D software products for digital city mapping, again with the help of local government investments. The BBS and the Aldin together resulted in a total customer base of 100,000. In October 2004, EHEnet and Search Partner signed the strategic cooperation agreement to jointly plan the creation of Homee, China's foremost property related search engine.

The firm grew a lot in the last years and now has over 400 employees and has attracted domestic venture capital. Zhejiang HeYi currently has more than 100,000 registered members with the website's daily average traffic surpassing 1 million people. Since its operation, the company has obtained consistent high praises from top level management and executive leaders in the retail trade. In 2005, the company was awarded the Zhejiang Top Ten Ecommerce Website and Hangzhou Ecommerce Application Enterprise.

Alliances of Zhejiang HeYi

In total Zhejiang HeYi has over 175 business partners: 100 in Hangzhou, 45 in Shanghai, 10 in Qingdao, 12 in Shenzhen, 2 in Xi'an, 2 in Changsha, 4 in Chengdu, 12 in Guangzhou, 1 in Weihai, 2 in Haerbin, and 3 in Xiamen.

The CEO of the company indicated that there are basically two types of alliances that matter for the success of his firm. One aspect is to establish relationship with the government, including the Provincial Economic and

Trade Commission, the Provincial Department of Information Industry, and the Development and Planning Commission. The government is now promoting the development of the hi-tech industry and informatization, and they take advantage of this to get funds and support from the government, such as in acquiring two of their partner firms. How do they build relationships with the government? Via friends, via direct visits and presenting products but also using the Chinese way: treating officials to dinner or tea and other gifts. Furthermore, Zhejiang HeYi will help internal departments like marketing and HRM to get in contact with government people working in the same areas to facilitate further integration between the government-firm alliance. For instance, as indicated by the CEO: "we introduce people from the marketing department to people working in the press, people in the BBS department to people working in Internet Supervision Bureau."

Another type of alliance fosters relationships with the companies in the same industry. CEOs join a club called WEB2 and meet every month, talking about technology, management and cooperation. Especially if the sites match up, there is a good basis for cooperation. For instance, as indicated by the CEO: "we cooperated with KouWei and another 100 websites in Hangzhou, both in reciprocal links and bundling of products. The general idea is that it is beneficial for all to cooperate and create a "brand store". Such a brand store of strategic alliance partners, formalized and visualized on each partner's website, creates legitimacy for the members and their products/services. The club serves managers to get away from their daily affairs and think at a higher level. People in the club are all entrepreneurs. They brainstorm every time they meet:

> The IT industry has grown for less than 10 years and the established relationships are quite simple. IT industry is still a young industry, and there is not much knowledge accumulation, especially in managerial experience. Apart from exploring by yourself, you should also learn from each other. Specifically, my firm's transformation was a result of learning from the BLOG of another company's CEO. Moreover, you can get contact with VC through social network.

Strategic alliances are crucial for the development of Zhejiang HeYi, in particular, the development of partnerships with local domestic firms and government agencies.

AnyRuan Case

AnyRuan was set up in 2002 in Hangzhou as the first mobile phone games software producer in China. It mainly provides mobile system wireless communication value added services: (1) mobile phone text messaging platform system software; (2) Cell phone MMS platform; (3) Mobile gaming software; (4) Internet-based information services (such as large file downloads, online games); and (5) MMS platform-basis operating and positioning systems. The headquarters are in Hangzhou, but they have offices in Beijing, Zhangsha, Lanzhou, Shenyang, Chengdu, Wuhan, Guangzhou, etc. The goal of the firm is to attract all CMDA 1X and GPRS users. In 2004 they entered the Thai market and they expect to expand into South East Asia, Europe and the US, which are markets in development. In 2007, Hangzhou AnyRuan received the "Zhejiang Technology Development of Award" for their "Digitalism Product Inner System Soft Platform."

Alliances of AnyRuan

"Every kind of alliances is a key factor to gain market access." The CEO has a clear view on the role of strategic alliances for his firm. The firm has alliances with almost all possible actors: state, business, individuals. For instance, they formed a value alliance with business leaders and competitors like Tupian and Lingsheng. They have a cooperation agreement with Zhejiang University for technology development, commercializing academic research results into products. After some time they developed a multimedia core technology, high speed development platform for mobile phone game development, data storage technology, and all have been breakthrough technologies, first appreciated and used by China Unicom users.

Cooperation with the government mainly reflects today's industry norms. Government approval decides whether or not products can be put on the market. Because the market does not operate standards, many matters are approved by the government: *"having guanxi is the same as having a market, firms requirements for new technology are not big, just fix government guanxi."*

Other business partners include large state-owned telecom operators: China Mobile, China Unicom, China Telecom business. They mainly cooperate with their business departments. The firm uses the partners channels, make embedded games for mobile phone clients. The firm has advantages of using the partners' channels to distribute the product. The competition in the wireless communication value added services is extremely fierce. This reflects today's telecom business market monopoly. Furthermore, the cooperation with the telecom partners is not transparent, the business rules are not standard, and cooperation relies mostly on secret chats, Telecom operators themselves formulate policy regulations. An example of a good relationship is China Unicom. The firm managed to establish a partnership with China Unicom through the government and personal relationships. It communicates with China Unicom every day in order to keep the partnership. As close partners, their market can be broadened through collaboration. The firm harmonizes resources and relationships through a proper business model, good personal relationship, cooperation between core team members and so on.

Strategic alliances are also crucial for the development of AnyRuan. The firm has a broad portfolio of alliances with almost all possible actors including the state, business, individuals, and large state-owned telecom operators.

DISCUSSION AND CONCLUSION: INSTITUTIONAL ALLIANCE CAPABILITY?

The survey of the literature resulted in five insights:

- alliances in China is not a niche subject and is published in general management journals;
- limited use of notion of alliance capability in studying alliances between Chinese firms;
- most interorganizational relationships fall in the category joint venture, strategic alliances or customer-supplier relations;
- majority of the papers deals with high-tech and manufacturing sectors; and
- majority of papers employs a survey method in which existing theories and hypotheses are tested within the Chinese context.

The case studies provide the following insights:

- Zhejiang HeYi: Two types of alliance partners: government and domestic private firms. The case suggests the importance of aligning and leveraging two sets of strategic alliances, with business and government. An institutional alliance secures information and insights into local policy changes and opportunities for investment. Business alliances are necessary for mitigating the risks of newness, that is, relative inexperience of the whole market; and
- AnyRuan: Two dominant types of alliance partners: state-owned enterprises and government bureaus. The case suggests the importance of such institutional alliances for access to market. Further, the case emphasized the need for a variety of partners, including local domestic firms and universities for technology development.

Interestingly, both firms are small private firms but managed to build strategic alliances with institutional partners. This raises the question as to how small private firms develop an alliance capability that incorporates both business and institutional partners.

Given the above findings and insights we suggest that firms in emerging economies may develop alternative strategies to secure appropriability and mobilize resources via strategic alliances and develop alliance capabilities that incorporate an *institutional dimension*. Emerging economies such as China undergo substantial institutional transformation (Peng, 2003). As opposed to reducing risk and uncertainty, institutions become a source of uncertainty, particularly regarding the formal protection of intellectual property, limitations in factor markets and higher search costs for critical resources (Peng, 2003; Wright, Filatotchev, Hoskisson, & Peng, 2005). In particular, we may summarize the institutional effect of China's transition economy on innovation along two dimensions: change and variety.

First of all, as a result of gradual institutional change, the co-existence of market and state socialist institutions may also leave some institutions ineffective or inert, obstructing the establishment and implementation of new institutions. The dismantling of state socialist institutions and the emergence of and experimentation with market institutions causes considerable uncertainty. The country resource environment is then left with an ambiguous and potentially conflicting set of rules to govern access to resources. The inefficiency of institutions—or "institutional baggage"—as a result of institutional change influences the accessibility of potential resources for firms to use (Qian, 2000; Roth & Kostova, 2003). Second, China has a variety of local institutions. The heterogeneity of China's local business environments is a consequence of China's decentralized government (Zheng, 2007). Local institutional variety causes wide differences in local resource bases and local governance mechanisms and increases horizontal competition of resources across localities (Dougherty & McGuckin, 2008). The consequence is that it is difficult to develop boundary-spanning innovation networks, tap into distant resource markets and accumulate and employ experience and generic skills across China. The diversity of local business systems causes significant variation in the availability and accessibility of resources across different business systems, as also evidenced by a relatively unstable and decentralized standardization regime for innovation (Van de Kaa, Greeven, & van Puijenbroek, 2013). Not only do different local institutions provide a diverging set of facilitative resources (e.g., different amount and quality of educational resources) but the enforcement of constraints on opportunism for firms also varies.

Our cases illustrate the point of developing alliances with a variety of partners but also identify an important role of developing strategic alliances with government bureaus and state-owned enterprises. Therefore, we suggest that developing a strategic alliance capability in China may need to consider institutional alliances as well as business alliances. In short, we suggest that an *institutional alliance capability* may be crucial in creating a competitive advantage for innovating firms in China's transition economy.

We suggest that further research may:

1. expand research on alliance capability in the context of Chinese-Chinese alliances;
2. explore how Chinese private firms develop alliances with state-owned firms and government bureaus; to what extent and how do firms develop alliances with institutional partners, i.e. institutional alliance capability;
3. move away from research on Chinese-Foreign joint ventures, shifting attention to Chinese alliances in the significant domestic private sector;
4. incorporate the institutional effects of an emerging economy on the development of alliance capability;
5. expand research on alliance capability beyond manufacturing and high-tech sectors but also include newly emerging sectors such as telecom and biotechnology; and
6. employ more explorative case methodology in acknowledgment of the rich and diverse context of alliances in China's emerging economy.

REFERENCES

Ahlstrom, D., & Bruton, G. D. (2010). Rapid institutional shifts and the co-evolution of entrepreneurial firms in transition economies. *Entrepreneurship Theory and Practice, 34*, 531-554.

Anand, B., & Khanna, T. (2000). Do firms learn to create value: The case of alliances. *Strategic Management Journal, 21*, 295-315.

Dougherty, S. M., & McGuckin, R. H. (2008). Local governments as industrial firms: An organizational analysis of China's transitional economy. *American Journal of Sociology, 10*, 263-301.

Draulans, J., de Man, A.-P., & Volberda, H. W. (2003). Building alliance capability: Management techniques for superior alliance performance. *Long Range Planning, 36*, 151-166.

Eisenhardt, K., & Schoonhoven, C. (1996). Resource-based view of strategic alliance formation: Strategic and social effects in entrepreneurial firms. *Organization Science, 7*, 136-150.

Gulati, R. (1998). Alliances and networks. *Strategic Management Journal, 19*, 293-317.

Hagedoorn, J. (1900). Organizational modes of inter-firm co-operation and technology transfer. *Technovation, 10*, 17-30.

Hamel, G. (1991). Competition for competence and inter-partner learning within international strategic alliances. *Strategic Management Journal, 12*(Special Issue: Global Strategy), 83-103.

Hamel, G., Doz, Y., & Prahalad, C. (1989). Collaborate with your competitors and win. *Harvard Business Review, 67*(1), 133-139.

Heimeriks, K. H., & Duysters, G. (2007). Alliance capability as a mediator between experience and alliance performance: An empirical investigation into the alliance capability development process. *Journal of Management Studies, 44*, 25-49.

Hitt, M. A., Ahlstrom, D., Dacin, M. T., Levitas, E., & Svobodina, L. (2004). The institutional effects on strategic alliance partner selection in transition economies: China vs. Russia. *Organization Science, 15*, 173-185.

Kale, P., Dyer, J. H., & Singh, H. (2002). Alliance capability, stock market response, and long-term alliance success: The role of the alliance function. *Strategic Management Journal, 23*, 747-767.

Kale P., & Singh, H. (1999). *Alliance capability and success: A knowledge-based approach*. Working paper, Wharton School, University of Pennsylvania.

Kale, P., Singh, H., & Perlmutter, H. (2000). Learning and protection of proprietary assets in strategic alliances: Building relational capital. *Strategic Management Journal, 21*, 217-237.

Kogut, B. (1988). Joint ventures: Theoretical and empirical perspectives. *Strategic Management Journal, 9*, 319-332.

Lambe, C. J., Spekman, R. E., & Hunt, S. D. (2002). Alliance competence, resources, and alliance success: Conceptualization, measurement, and initial test. *Journal of the Academy of Marketing Science, 30*(2), 141-158.

Murray, J. Y., Kotabe, M., & Zhou, J. N. (2005). Strategic alliance-based sourcing and market performance: Evidence from foreign firms operating in China. *Journal of International Business Studies, 36*, 187-208.

Peng, M. W. (2003). Institutional transitions and strategic choices. *Academy of Management Review, 28*, 275-296.

Perks, H., Khan, K., & Zhang, C. (2009). An empirical evaluation of R&D-marketing NPD integration in Chinese firms: The *guanxi* effect. *Journal of Product Innovation Management, 26*, 640-651.

Porter, M. E., & Fuller, M. B. (1986). Coalitions and global strategy. In M. E. Porter (Ed.), *Competition in global industries* (pp. 315-343). Boston, MA: Harvard Business University Press.

Powell, W. W., Koput, K., & Smith-Doerr, L. (1996). Inter-organizational collaboration and the locus of innovation: Networks of learning in biotechnology. *Administrative Science Quarterly, 41*, 116-145.

Qian, Y. Y. (2000). The process of China's market transition (1978-1998): The evolutionary, historical, and comparative perspectives. *Journal of Institutional and Theoretical Economics, 156*(1), 151-171.

Roth, K., & Kostova, T. (2003). Organizational coping with institutional upheaval in transition economies. *Journal of World Business, 38*, 314-330.

Sanchez, R., Heene, A., & Thomas, H. (1996). Dynamics of competence-based competition: Theory and practice in the new strategic management. Oxford, UK: Elsevier Pergamon.

Siu, W. S., & Bao, Q. (2008). Network strategies of small chinese high-technology firms: A quantitative study. *Journal of Product Innovation Management, 25*, 79-102.

Tsui, A. S., Bian, Y., & Cheng, L. (2006). *China's domestic private firms: Multidisciplinary perspectives on management and performance*. London, UK: M. E. Sharpe.

Van de Kaa, G., Greeven, M., & van Puijenbroek, G. (2013). Standards battles in China: Opening up the black-box of the Chinese government. *Technology Analysis & Strategic Management, 25*, 567-581.

World Bank. (2012). *China 2030—Building a modern, harmonious, and creative high-income society*, The World Bank, Development Research Centre of the State Council, The People's Republic of China, Conference edition.

Wright, M., Filatotchev , I., Hoskisson, R. E. E., & Peng, M. W. (2005). Strategy research in emerging economies: Challenging the conventional wisdom. *Journal of Management Studies, 42*, 1-33.

Xin, K. & Pearce, J. L. (1996). Guanxi: Connections as substitutes for formal institutional support, *Academy of Management Journal, 39*, 1641-1658.

Zheng, Y. (2007). *De facto federalism in China: Reforms and dynamics of central-local relations*. London, UK: World Scientific.

CHAPTER 11

THE DISTINCT IMPACTS OF PUBLIC-PRIVATE ALLIANCES VERSUS PRIVATE-PRIVATE ALLIANCES ON PRODUCT INNOVATION

An Empirical Assessment

Jie Wu and Nitin Pangarkar

ABSTRACT

In this study we unbundle strategic alliances into two distinct categories: private-private strategic alliances (interfirm strategic alliances) and public-private strategic alliances (strategic alliances between firms and public organizations). Drawing on Das and Teng's (2000a) seminal work, we argue that the two forms of strategic alliances differ in terms of their performance implications specifically in terms of their impact on innovation output. We submit that private-private alliances will have a positive impact on innovation output while public-private alliances will have a negative impact. Results from a panel study of 851 Chinese firms across five manufacturing industries provide strong support for our predictions.

Strategic Alliances for Innovation and R&D, pp. 249–268

INTRODUCTION

Strategic alliances refer to voluntary arrangements between independent organizations to share resources with an aim to achieve mutual strategic objectives (Das & Teng, 2000a). Though this definition does not exclude strategic alliances between commercial, for-profit firms and nonprofit or public organizations (called "public-private" strategic alliances in this chapter), the extensive literature on alliances has often focused on alliances where all partners were profit-motivated (called "private-private" strategic alliances in this chapter) (Arora, Fosfuri, & Gambardella, 2001; Das & Teng, 2000a; Wu, 2014). This exclusion is surprising since public-private alliances might account for 15% or more of the total number of alliances. The importance of public-private alliances can also be seen from a quote from the German pharmaceutical company Boehringer Ingelheim, among the most admired pharmaceutical companies in the world:

> We continue to support cooperations with academic research scientists to better define the underlying causes and molecular mechanisms of human disease. This is part of our involvement in public-private partnerships and guided by the notion of active citizenship and service to society. (Boehringer Ingelheim, 2011)

Strategic alliances are formed for various purposes including exchanging technological know-how, entering new geographic markets, developing new products, and enhancing marketing strategies, among others (Ahuja, 2000a; Deeds & Hill, 1996; Wu, 2012). In this study we focused our attention on strategic alliances that are formed with the aim to innovate. In an increasingly competitive world, innovation has become a competitive necessity and firms often forge alliances to boost their innovation output. R&D or technology alliances (Powell, Koput, & Smith-Doerr, 1996), whether in the home country or abroad, allow partners to share R&D costs, pool risks, and enjoy access to firm-specific know-how and commercialization resources (Arora et al., 2001; Hagedoorn, Link, & Vonortas, 2000). According to one estimate, technology alliances represent as much as 23% of all types of alliances in North America compared with 14% in Western Europe and 12% in Asia (Kang & Sakai, 2000). Technology alliances, themselves, can take many forms, including joint research activities, technology co-development, contract research, and technology exchange including licensing and cross-licensing (National Science Board, 2002).

We propose and empirically demonstrate that not all technology alliances are "equal." especially with regard to their impact on innovation. In fact, we propose that the two distinct forms of strategic alliances (public-private and private-private) have contrasting influence on product innovation. We tested our hypotheses using a panel study of 851 Chinese manufacturing firms across five manufacturing industries.

THEORY DEVELOPMENT AND HYPOTHESES

The Resource-Based View and Strategic Alliances

The resource-based view (RBV) argues that bundles of resources are heterogeneously distributed across firms and this heterogeneity persists over time. Idiosyncratic resources that are difficult to imitate and simultaneously create value will help a firm achieve competitive advantage and sustain it over time (Barney, 1991; Dierickx & Cool, 1989; Wernerfelt, 1984). Alliance scholars have widely acknowledged that strategic alliances are important vehicles to "aggregate, share, or exchange valuable resources with other firms when these resources cannot be efficiently obtained through market exchange or mergers/acquisitions (M&As)" (Das & Teng, 2000a, p. 37).

In addition to accessing a partner's resources in the short term, strategic alliances can also help firms "retain" their own valuable resources for future use. Firms sometimes need to engage in alliances to prevent their know-how from decaying (Kogut, 1988). Specifically, strategic alliances can help them "remember by doing" rather than "forget by not doing." This may be especially useful in situations where firms have resources that may be underutilized in the short term and hence subject to the risk of decaying. For example, a firm may have excess research personnel who don't have meaningful research projects to work on. Rather than laying these personnel off (and possibly have them join competitors), the firm can perform contract research services for its alliance partners or join hands with another partner in a commercialization venture where the firm performs the research and its partner takes charge of marketing (Das & Teng, 2000a). By doing so, not only will the firm retain its expertise, but it will also have an opportunity to make additional profits should its alliance succeed.

Public-Private Versus Private-Private Alliances

Public-private alliances which can be exemplified by contractual relationships between a firm and a local university, or between a firm and a government research institute are, in fact, quite common (George, Zahra, & Wood, 2002; Wu, 2011). They are undertaken with a variety of motives to leverage social benefits from public-supported activity (e.g., government sponsored R&D resources given to universities and other public institutions see, e.g., O'Brien, Wakeham, & Walsh, 1999; OECD, 2004), to enhance the competitive position of the industry in the global marketplace (Rausser, Simon, & Ameden, 2000), and to leverage industrial R&D to meet defense or government needs (Link, 2002). Between 1998 and 2000, 8% of innovative firms within EU cooperated with a University and another 8% cooperated with a research institute (OECD, 2004). According to the National Science Board (2002), for the 1985-2000 period, U.S. universities participated in 15% of technology joint ventures, and 11% of the alliances had at least one federal laboratory member. All these statistics point towards the prevalence of public-private alliances. Since private-private alliances have been discussed extensively by prior studies (Baum, Calabrese, & Silverman, 2000; Powell et al., 1996), we will not discuss them except in terms of contrasting them with public-private alliances.

There are similarities as well as differences between public-private and private-private alliances. At a broad level, both types of alliances bring together complementary resources and hence have the potential to create synergy (Baum et al., 2000; Pangarkar & Klein, 2001, 2004; Powell et al., 1996). We believe that it is difficult to make a general remark about which type of alliances will have greater potential synergies. It is, however, possible to make specific predictions in other respects such as which type of alliances will have greater operational conflicts.

Below, we argue that the operational conflicts in public-private alliances will be greater. Since they bring together two or more different organizations, cultural differences are likely to create conflicts and frictions in both types of alliances (Bucklin & Sengupta, 1993; Hardy & Phillips, 1998; Khanna, Gulati, & Nohria, 1998). Partner conflicts can be classified into two types: competing interests and operational conflicts. Interest conflicts largely relate to competing interests of alliances partners attributable to incompatible goals, disagreements regarding resource allocation, and opportunistic behavior (Pangarkar & Choo, 2001). Stopford and Wells (1972) discussed the conflicts between a multinational firm and its local partner regarding control. The other type of partner conflict relates to the diversity of the partners' strategic orientations, managerial practices, and organizational culture that creates challenges for cooperative activities (Das & Teng, 2000a; Olk, 1997). When two partners are very different in terms of their organizational cultures, managerial practices, strategic orientations and technological systems, they cannot work very well and coordination costs will increase in direct proportion to the magnitude of these differences (Harrigan, 1988; Park & Ungson, 1997; Whetten, 1981). While both interest conflicts and operational conflicts can (and probably do) occur in both types (public-private as well as private-private) of strategic alliances, the former is more often associated with private-private strategic alliances and the latter with public-private strategic alliances (Bucklin & Sengupta, 1993; Zaheer, McEvily, & Perrone, 1998). For example, in a firm-university alliance, while a firm may be more interested in commercializing new technology by launching novel products, a university (and its researchers) may be more interested in basic research, publications and their impact factors. With its public identity, a public organization (e.g., a university or a government research institute) is also required to get involved in public activities to live up to its public responsibilities and would be reluctant to choose commercial considerations over these responsibilities (Pearce, 1997). Even if partners actively try to address these conflicts (Thomas & Trevino, 1993), these efforts constitute the "non-value-enhancing activities" that make the alliance less productive, leading to reduced alliance performance such as innovation output (Zaheer et al., 1998, p. 146). In summary, we believe that the operational conflicts in public-private alliances are likely to take a greater toll of the innovation of the alliance. In this regard, Rothaermel and Deeds (2006) made the following comment:

> alliances with universities and other research institutions would require the greatest amount of a biotechnology firm's alliance management capability since the difference between non-profit and for-profit institutions tend to be more fundamental than differences among various types of for-profit institutions. (p. 435)

They further noted that though research universities are increasing their commercial involvement (Thursby & Thursby, 2002), they are at an early stage of developing this mindset (Bowie, 1994; Di Gregorio & Shane, 2003) and hence operational conflicts are quite likely in public-private alliances.

There are additional differences across these two types of alliances with regard to the achievement of scale economies, risks of partner opportunism and availability of subsidized (or sometimes low-cost) resources. Turning first to the scale economies, the domain of partners is likely to be overlapping to a greater extent in private-private alliances than public-private alliances, which implies that the potential to achieve scale economies is greater in the former type than the latter type. When a firm collaborates with another firm to develop technology, each partner can potentially receive greater returns (in terms of knowledge and/or funds) than if it were to undertake a similar project independently (Ahuja, 2000a). To illustrate, if each of two partners in a technology alliance contributes an equal amount of dollars (e.g., $X) to the alliance, then the alliance will have double amount of dollars (e.g., $2X) available to it, in addition to any internal resources allocated by each partner. The benefits of scale economies and risk sharing in research arise when larger projects generate significantly greater quantum of knowledge (or more valuable knowledge) than smaller projects. As Ahuja (2000a, p. 429) noted, "If the transformation of technology is characterized by increasing returns, such an investment will lead to a more than proportionate return in terms of innovation output, benefiting both firms significantly." As argued earlier, because of overlapping domains, a private-private alliance is more likely to enable partner firms to take advantage of such scale economies than a public-private alliance (Hill & Hellriegel, 1994; Wu, 2011).

Public-private alliances enable private firms to enjoy a "subsidy" because the research of many public organizations is partly or fully funded by the government. In addition to direct funding, public organizations might also have resources such as equipment and manpower that have been acquired (or are being compensated at) below the market price. In fact, Cohen, Nelson, and Walsh (2002) found that, despite the challenges posed by a lack of profit orientation of public organizations, more than one-third of their respondent firms used public research to address existing (applied) problems. Their results suggest that the perception that public-private alliances are more suitable for basic research and solving long-term problems may be an (inaccurate) stereotype, possibly because cost savings for the private partner would induce use of public-private alliances for a broader set of purposes. Rothaermel and Deeds (2006) cite the examples of a paper company and the pharmaceutical company Eli Lilly using alliances with public organizations to solve immediate research challenges. In summary, to achieve a certain output, public-private alliances might need lesser resources than other alliances.

Implications of the Type of Alliance for Product Innovation

We submit that private-private strategic alliances will have a positive impact on product innovation output for at least three reasons. First, as we noted earlier, though conflicts characterize all alliances, the operational conflicts due to fundamental differences in orientation (commercial/profit seeking versus broader objective function for the public partner) will hamper the innovation output of public-private alliances. Second, because of overlapping domains, private-private alliances have greater opportunities to achieve scale economies. Thirdly, simply because of their sheer number (roughly 80% of all alliances tend to be private-private alliances), any partner is likely to have had greater experience in managing private-private alliances. Having greater experience will also imply that firms will be more adept at achieving desired outcomes such as innovation from these alliances than with public-private alliances.

On the flip side, private-private strategic alliances may be impacted by opportunism by partners. Though both firms may be interested in profit maximization, they may not pursue joint profit maximization but pursue private benefits, which gives rise to potential conflicts of interests in alliances (Khanna et al., 1998). Pursuit of private benefits may lead to less common benefits and, consequently, to unsatisfactory alliance performance, such as an alliance breaking apart without any innovation (Park & Russo, 1996). The risks of opportunism can, however, be mitigated through adoption of appropriate strategies before the formation of the alliance (e.g., choosing a reputable and trustworthy partner) as well as during the operation of the alliance (e.g., through continuous communication with the partner and/or effective monitoring) (Medcof, 1997; Pangarkar & Klein, 2004).

In summary, while both types of alliances have their benefits and drawbacks (see Table 11.1 for a summary of the arguments advanced so far), we submit that positive impact of private-private strategic alliances will outweigh the obstacles in these alliances, leading to better innovation performance (Das & Teng, 2000b). Operational conflicts in the public-private alliances, on the other hand, might overwhelm some of the synergistic effects and lead to lower innovation. Hence, we propose the following two hypotheses:

Hypothesis 1a: *Private-private strategic alliances have a positive impact on the product innovation output of the focal firm.*

Hypothesis 1b: *Public-private strategic alliances have a negative impact on the product innovation output of the focal firm.*

METHODS

Data and Sampling

We used data from a large survey conducted by the Enterprise Survey Organization of China's National Bureau of Statistics in collaboration with the World Bank under a project entitled Competitiveness, Technology and Firm Linkages. The survey covered 1,500 firms from five cities in China: Beijing, Chengdu, Guangzhou, Shanghai and Tianjin. Three hundred firms from each city were selected by experts with an aim to achieve a representative sample given predetermined distributions by city, industry,

Table 11.1. Characteristics of the Different Types of Alliances

Both Public-Private and Private-Private Alliances	*Public-Private Alliances*	*Private-Private Alliances*
Resource complementarity leading to better performance including innovation output	Lower learning because of lower frequency in general as well as by the focal firm	Greater learning benefits due to greater frequency of formation in general as well as by the focal firm (vicarious learning as well as learning by doing)
Culture differences and frictions	Subsidies enjoyed by the public partner because of government funding as well as cheaper input costs (equipment and/ or manpower)	Commonality in partners' goals of successful commercialization and greater profits → less operational conflicts
	Operational conflicts because of fundamental differences in the way in which public and private organizations operate	Risks of opportunism by partners which may be mitigated by following appropriate strategies before the formation of the alliance and/or during the operation of the alliance

and size. For a more detailed description of the data, see Hallward-Driemeier, Wallsten, and Xu (2003).

The survey administrators employed a rigorous methodology for administering the questionnaire, including adopting the following procedures. The questionnaire was prepared in English and translated into Chinese by scholars competent in both languages and with substantial research experience. A translation-back-translation technique was employed to make sure that there were no misunderstandings. Before the formal survey, the questionnaire was pilot tested in interviews with 17 business managers to determine whether the questions were properly worded and well understood in the context of Chinese business. A letter of introduction was then hand-delivered to the chief executive officer (CEO)/general manager of each company, explaining the purpose of the study and inviting participation. The top executives (general managers, CEOs or other senior managers) were then contacted by telephone within 2 weeks to ensure their participation in the study and to make appointments for onsite interviews. This rigorous survey method ensured access to the appropriate respondents, confirmed the correct use and understanding of the terms, and provided high response rates (Chen & Wu, 2011). Each completed questionnaire was immediately checked by a supervisor to determine whether it had been filled out in accordance with the instructions.

All questionnaires were verified by supervisors before data entry (for more details, see World Bank, 2003).

The survey comprised two separate questionnaires answered by two different groups of respondents. One was completed by accountants or personnel managers who provided basic profile information such as about a firm's ownership and revenues, research and development (R&D) expenditures and labor force size. The other was completed by senior line managers (e.g., head of manufacturing or a general manager) who provided information on innovation outcomes. The sample covered ten industry sectors, including five manufacturing sectors (electronic equipment, electronic components, consumer products, vehicles and vehicle parts, and apparel and leather goods) and five service sectors (accounting, advertising and marketing, business logistics, communications, and information technology). Because manufacturing and service industries exhibit salient differences (e.g., in their innovation patterns, Sirilli & Evangelista, 1998), this study used responses from the manufacturing firms only. After excluding items with missing values, the final sample comprised 851 firms.

To assess the risk of heteroscedasticity (a possibility because data was pooled across industries and cities), the panel data were analyzed using White's generalized test (Bowen & Wiersema, 1999). The Breusch-Pagan test statistics suggested that heteroscedasticity is not a concern ($\chi^2 = 16.92$, $p = 0.37$). The estimated residuals were also plotted against the independent variables and no systematic patterns of heteroscedasticity were found (Wooldridge, 2009). We also created dummy variables representing industries and locations to model coefficient variations, thus alleviating potential concerns about heteroscedasticity associated with pooling of the data (Greene, 1994).

To overcome common method bias, we used the information provided by two different sets of respondents providing information at different points in time. Specifically, the information on the dependent and independent variables was provided by different respondents from the same firm. This research design substantially minimizes the risk of common method variance (Eisenhardt & Tabrizi, 1995). In addition, a Harmon one-factor test for common method bias showed that the first factor explained less than 41% of the variance, indicating that common method bias was not a serious concern in this study.

Measures

Formation of private-private strategic alliances (Private) took a value of 1 if the firm had formed at least one technological strategic alliance in the years 1999 or 2000, and 0 otherwise (Ahuja, 2000a; Wu, in press). We verified alliance formation information from survey responses by checking firm websites, financial newspapers, and written reports published for a randomly selected sample of firms. The cross-checking bolstered our confidence in the accuracy of the data.

Public-Private Strategic Alliances (Public) took a value of 1 if the firm had formed at least one technological strategic alliance with prominent public organizations in the years 1999 or 2000, and 0 otherwise. Prominent public organizations included renowned universities, government research institutes and government agencies. We employed cross-checking approaches as described above and did not find any evidence of inaccuracies in the data.

Product Innovation was measured by the number of new products a firm had introduced to the market in 2000. The number of new products is an important indicator of product innovation because it "indicates the potential commercial significance of a firm's innovation activities" and "most innovations cannot influence firm performance until the ideas have been put into use and introduced to the market" (Katila, 2002, p. 996). Prior studies have shown that the introduction of new products increases market share and market value (Chaney & Devinney, 1992), improves firm performance (Roberts, 1999), and enhances a firm's survival chances (Banbury & Mitchell, 1995).

We included several control variables that could be alternative explanations of innovation output. We controlled for a firm's *foreign ownership* because a high level of foreign ownership will motivate a foreign partner to actively engage in the management of the focal firm in terms of sharing useful knowledge and information, thus boosting product innovation (Lyles & Salk, 1996). It was measured by aggregating the stake by foreign investors including foreign individuals, foreign institutional investors, foreign firms, and foreign banks. Second, because industry import penetration might affect a firm's incentive to perform innovation (Bowen & Wiersema, 2005), we controlled *industry import penetration* measured as the ratio of imports to total domestic sales for the main industry of the focal firm. Third, we included *firm performance* to control for firm heterogeneity in terms of financial resources to invest in new product development. Firm performance was measured by return on sales (ROS). Fourth, because large firms have more resources to devote to product innovation, we controlled for *firm size,* measured as the logarithm of the number of employees (Wu & Chen, 2012). We also controlled for *firm age* as older firms may suffer from core rigidities or inertial tendencies and hence have lower innovation output. Firm age was measured as the number of years since a firm's founding (Wu, 2012). In addition, we controlled for whether a firm is a *publicly listed company* (1 = yes; 0 = no) as private firms lack access to capital markets in China (Li & Atuahene-Gima,

2001). We also controlled for the total number of firms in the industry, a proxy for competitive intensity which could impact the incentives to undertake innovation and, consequently, innovation output. Also, because our sample includes firms from five manufacturing industries, we created four *industry dummy* variables using the apparel and leather goods industry as the base group. Additionally, we included four *city dummy* variables to control location effect, since the sampling firms are drawn from five Chinese cities. We used Shanghai as the base group in the analysis.

Statistical Modeling

All models were estimated using robust maximum likelihood estimation (MLE). The dependent variable, the number of new products introduced, was a count variable. Poisson or negative binomial regression modeling can be used with count data, but a Poisson regression model requires that the mean be equal to the standard deviation, so it was not suitable for this dataset and negative binomial regression was applied in the analyses. The negative binomial regression model is:

$$y_j \sim \text{Poisson}(\mu_j^*),$$

$$\text{where } \mu_j^* = \exp(\mathbf{x}_j\beta + \text{offset}_j + v_j),\ e^{v_j} \sim Gamma(1/\alpha, \alpha).$$

e^{v_j} follows a gamma distribution with expectation ab and variance ab^2, α is the overdispersion parameter. The larger the value of α, the greater the overdispersion (Long & Freese, 2006). Although time subscripts have been omitted to reduce the complexity of the above equation, all explanatory variables were lagged one year, taking into consideration a possible delay before the effects of explanatory variables would be reflected in product innovation outcomes.

Although this study used strategic alliances as predictor variables, not all the firms have the same chance of forming alliances. As a result, firms that form strategic alliances may differ in other important ways from those which do not. Heckman's selection model was used to account for this self-selection effect (Heckman, 1979; Shaver, 1998). The models were evaluated in two stages. In the first stage, probit regression was used to estimate the probability that a firm formed strategic alliances as a function of firm age, firm size, foreign ownership percentage and the industries dummies. The predicted value derived from the first stage was then transformed into an inverse Mills ratio. The inverse Mills ratio was then included as a regressor in the second stage model to estimate the probability of new product introduction (Heckman, 1979; Shaver, 1998).

This two-stage procedure tends to generate consistent and asymptotically efficient estimates (Heckman, 1979).

RESULTS

In Table 11.2, we report the descriptive statistics for the variables. A review of the correlations among the independent variables suggests that multicollinearity is not a major concern. This was confirmed with an analysis of variance inflation factors (VIF). The VIF values range from 1.16 to 4.71, well below the cutoff threshold of ten, which indicates that there are no serious multicollinearity problems in the models (Hair, Anderson, Tatham, & Black, 1998).

In Table 11.3, we provide the estimation results for testing Hypotheses 1a and 1b. M1 includes the controls, M2 and M3 add each of the key variables (public-private alliances and private-private alliances) in turn and M4 includes both the key variables. The log-likelihood ratios and chi-squares statistics are better for models M2 to M4 suggesting the addition of the key variables improves the fit of the models.

Under Hypothesis 1a we had predicted a positive relationship between private-private strategic alliances and product innovation success. The coefficient of Private in M2 and M4 is positive and significant ($\beta = 0.40$, $p \leq 0.001$ and $\beta = 0.61, p \leq 0.001$) indicating that private-private strategic alliances positively impact product innovation. Hence, Hypothesis 1a was supported.

Under Hypothesis 1b, we had predicted a negative relationship between public-private strategic alliances and product innovation. As shown in M3, the coefficient of the variable is positive and significant ($\beta = -0.22$, $p \leq 0.01$; $\beta = -0.42, p \leq 0.001$, respectively), indicating that public-private strategic alliances negatively impact product innovation. Hence, Hypothesis 1b was also supported.

The robustness of these results was tested in several ways. First, to reduce any concerns that might arise because the sample included firms that did not launch any new products, the models were re-estimated with a sub-sample of 641 firms, all of whom had introduced at least one new product during the period studied. The results did not change in any substantial way. Furthermore, one could argue that the pattern of regulated industries (the transportation industry) may be different from that of unregulated industries such as electronic equipment, as the Chinese government exerts more constraints on the former (Wu, 2012). To address this potential concern, the transportation industry was excluded from the sample and all of the models were then reestimated. The new results were highly consistent with the ones reported.

Table 11.2. Means, Standard Deviations and Correlations

	Variables	*Mean*	*SD*	*1*	*2*	*3*	*4*	*5*	*6*	*7*	*8*	*9*	*10*
1	New product innovation	1.925	7.309	1.000									
2	Public strategic alliances	0.103	0.305	0.045	1.000								
3	Private strategic alliances	0.035	0.185	0.101*	0.312*	1.000							
4	Firm age	16.593	15.935	–0.044	0.095*	0.019	1.000						
5	Firm size	5.526	1.430	0.112*	0.246*	0.135*	0.342*	1.000					
6	Public listed	0.024	0.152	0.095*	0.100*	0.097*	–0.035	0.204*	1.000				
7	Firm performance	1.027	22.791	–0.012	–0.017	–0.010	–0.007	0.013	–0.009	1.000			
8	Industrial rivals	2.878	1.690	0.011	–0.011	0.049	0.070*	–0.062	0.010	0.014	1.000		
9	Foreign ownership	17.224	29.3[illegible]1	0.046	–0.075*	–0.072*	–0.265*	0.020	–0.044	0.064	–0.093*	1.000	
10	Import penetration	8.898	17.586	0.018	0.022	0.018	–0.044	0.015	–0.002	–0.019	–0.076*	0.098*	1.000

* indicates significance at the $p \leq 0.05$ level of confidence.

Table 11.3. Results of Hypotheses Testing

Variables	*M1*	*M2*	*M3*	*M4*
	Dependent variable: Innovation output (Number of new products			
Constant	–0.22	–0.05	–0.33	–0.18
	(0.25)	(0.25)	(0.26)	(0.25)
Firm age	–0.02***	–0.02***	–0.02***	–0.02***
	(0.00)	(0.00)	(0.00)	(0.00)
Firm size	0.20***	0.19***	0.22***	0.21***
	(0.02)	(0.02)	(0.02)	(0.02)
Pubic listed	0.33**	0.24*	0.37***	0.29**
	(0.10)	(0.11)	(0.11)	(0.11)
Firm performance	–0.01	–0.01	–0.02	–0.01
	(0.02)	(0.01)	(0.02)	(0.02)
Foreign ownership	0.11***	0.10***	0.10***	0.09***
	(0.02)	(0.02)	(0.02)	(0.02)
Industrial rivals	0.01***	0.01***	0.01***	0.01***
	(0.00)	(0.00)	(0.00)	(0.00)
Import penetration	0.00	0.00	0.00	0.00
	(0.00)	(0.00)	(0.00)	(0.00)
Inverse Mill's ratio	0.87	0.69	1.04*	0.87
	(0.46)	(0.46)	(0.46)	(0.47)
Industry dummy	(included)	(included)	(included)	(included)
Location dummy	(included)	(included)	(included)	(included)
Private-private alliances (Private)		0.40***		0.61***
		(0.09)		(0.10)
Public-private alliances (Public)			-0.22**	–0.42***
			(0.08)	(0.09)
Log-likelihood	–2,551.034	–2,542.256	–2,546.883	–2,529.768
AIC	5,140.069	5,124.511	5,133.766	5,101.536
BIC	5,230.251	5,219.440	5,228.694	5,201.211
Degrees of freedom (*df*)	16	17	17	18
χ^2	519.781	537.338	528.084	562.314
Prob. > $\chi 2$	0.000	0.000	0.000	0.000

Note: N= 851; Standard errors in parentheses

* $p < 0.05$, ** $p < 0.01$, *** $p < 0.001$ one tailed-test for the hypothesized variables and two-tailed test for the controls.

Among the control variables, firm size, public listed status, foreign ownership and higher competitive intensity of the industry all boosted innovation output. On the other hand, firm age impacted innovation negatively.

DISCUSSION

Building on the arguments of the RBV, this study examined the impact of two distinct types of strategic alliances public-private alliances and private-private alliances on the innovation output of Chinese manufacturing firms. Our conceptual arguments suggested that the two forms of strategic alliances impact innovation output differently, thus suggesting the need for a more nuanced view. Though we are in agreement with the prior literature about the importance of pooling and accessing resources through strategic alliances (Ahuja, 2000a; Das & Teng, 2000a; Wu, 2012), we believe that our nuanced view is more appropriate for explaining innovation, especially in an emerging economy context.

We found support for the argument that private-private alliances boost innovation in the short term through one or more of the following mechanisms: learning effects, complementarity of resources, scale economies in R&D and lower frictions. While public-private alliances may also benefit from complementarity of resources, they don't benefit from either learning effects or scale economies. Additionally, the potential of culture clashes in these alliances might reduce their innovation potential, at least in the short term. These factors might explain their observed negative impact.

In general, we found that Chinese firms with stronger resources (e.g., publicly listed firms, large-sized firms and firms with foreign ownership) have higher levels of innovation. Possession of resources may be especially critical in an emerging economy with immature institutions such as capital markets. We also found that higher competitive intensity of an industry leads to innovation supporting the view of some economists (e.g., Geroski, 1990) that lack of market power (or high competition) stimulates innovation.

We acknowledge a few limitations which, in turn, suggest interesting avenues for future research. First, we only looked at innovation in the short term and the effects of these different types of alliances on long term innovation may be quite different. The bolstering of knowledge base through public-private partnerships could have a positive influence on innovation in the long term, which would explain why prominent firms like Boehringer Ingelheim consider these alliances to be important for their strategy. Second, it is necessary to be cautious in inferring any causal relationships between the key constructs included in this study. The arguments tested are based on theoretical logic and the relationships were tested using panel data from one country. Future research using a longitudinal design is needed to confirm the existence of the causality proposed here. Although China shares many characteristics with other emerging market contexts, future studies could replicate our models using data from firms in other emerging markets to confirm the generalizability of the findings. This is especially important given the heterogeneous institutions of emerging markets, which will affect innovation by firms.

The results of this study raise several exciting questions for future research. Researchers might profitably explore how different types of strategic alliances affect firms' different degrees of novelty of innovation. For example, it would be interesting to examine whether private-private strategic alliances and public-private strategic alliances affect incremental and radical innovation differently. Furthermore, rather than their independent effect, future research could also examine how private-private strategic alliances interact with public-private alliances in impacting subsequent innovation output.

In conclusion, this study examined the effect of two distinct types of strategic alliances: public-private strategic alliances and private-private strategic alliance. We observed important differences in the impact of these two types of alliances which may be a function of the environment (specifically institutional characteristics) as well as fundamental differences in these types of alliances. Public organizations in China (and other emerging markets) may be at an earlier stage of evolution which could impact the innovation output of public-private alliances.

REFERENCES

Ahuja, G. (2000a). Collaboration networks, structural holes, and innovation: A longitudinal study. *Administrative Science Quarterly, 45*, 425-455.

Ahuja, G. (2000b). The duality of collaboration: Inducements and opportunities in the formation of interfirm linkages. *Strategic Management Journal, 21*, 317-343.

Arora, A., Fosfuri, A., & Gambardella, A. (2001). Markets for technology and their implications for corporate strategy. *Industrial and Corporate Change, 10*, 419-451.

Banbury, C. M., & Mitchell, W. (1995). The effect of introducing important incremental innovations on market share and business survival. *Strategic Management Journal, 16*(S1), 161-182.

Barney, J. B. (1991). Firm resources and sustainable competitive advantage. *Journal of Management, 17*, 99-120.

Baum, J. A., Calabrese, T., & Silverman, B. S. (2000). Don't go it alone: Alliance network composition and startups' performance in Canadian biotechnology. *Strategic Management Journal, 21*, 267-294.

Boehringer Ingelheim. (2011). *Annual report 2011*. Ingelheim am Rhein, Germany: Boehringer Ingelheim GmbH. Retrieved from http://www.boehringer-ingelheim.com/corporate_profile/annual_report.html

Bowen, H. P., & Wiersema, M. F. (1999). Matching method to paradigm in strategy research: Limitations of cross-sectional analysis and some methodological alternatives. *Strategic Management Journal, 20*, 625-636.

Bowen, H. P., & Wiersema, M. F. (2005). Foreign based competition and corporate diversification strategy. *Strategic Management Journal, 26*, 1153-1171.

Bowie, N. E. (1994). *University-business partnerships: An assessment*. Lanham, MD: Rowman & Littlefield.

Bucklin, L. P., & Sengupta, S. (1993). Organizing successful co-marketing alliances. *Journal of Marketing, 57*(2), 32-46.

Chaney, P. K., & Devinney, T. M. (1992). New product innovations and stock price performance. *Journal of Business Finance & Accounting, 19*, 677-695.

Chen, X., & Wu, J. (2011). Do different guanxi types affect capability building differently? A contingency view. *Industrial Marketing Management, 40*, 581-592.

Cohen, W. M., Nelson, R. R., & Walsh, J. P. (2002). Links and impacts: The influence of public research on industrial R&D. *Management Science, 48*, 1-23.

Das, T. K., & Teng, B. (2000a). A resource-based theory of strategic alliances. *Journal of Management, 26*, 31-61.

Das, T. K., & Teng, B. (2000b). Instabilities of strategic alliances: An internal tensions perspective. *Organization Science, 11*, 77-101.

Deeds, D. L., & Hill, C. W. L. (1996). Strategic alliances and the rate of new product development: An empirical study of entrepreneurial biotechnology firms. *Journal of Business Venturing, 11*, 41-55.

Di Gregorio, D., & Shane, S. (2003). Why do some universities generate more start-ups than others? *Research Policy, 32*, 209-227.

Dierickx, I., & Cool, K. (1989). Asset stock accumulation and sustainability of competitive advantage. *Management Science, 35*, 1504-1511.

Eisenhardt, K. M., & Tabrizi, B. N. (1995). Accelerating adaptive processes: Product innovation in the global computer industry. *Administrative Science Quarterly, 40*, 84-110.

George, G., Zahra, S. A., & Wood, D. R. (2002). The effects of business-university alliances on innovative output and financial performance: A study of publicly traded biotechnology companies. *Journal of Business Venturing, 17*, 577-609.

Geroski, P. A. (1990). Innovation, technological opportunity, and market-structure. *Oxford Economic Papers-New Series, 42*, 586-602.

Greene, W. H. (1994). Accounting for excess zeros and sample selection in Poisson and negative binomial regression models. *NYU Working Paper No. EC-94-10*, New York University, New York, NY.

Hagedoorn, J., Link, A. N., & Vonortas, N. S. (2000). Research partnerships. *Research Policy, 29*, 567-586.

Hair, A., Anderson, R. E., Tatham, R. L., & Black, W. C. (1998). *Multivariate data analysis*. Upper Saddle River, NJ: Prentice-Hall,.

Hallward-Driemeier, M., Wallsten, S., & Xu, L. C. (2003). *The investment climate and the firm: Firm-level evidence from China*. World Bank Policy Research Working Paper No. 3003. Available at SSRN: http://ssrn.com/abstract=373946.

Hardy, C., & Phillips, N. (1998). Strategies of engagement: Lessons from the critical examination of collaboration and conflict in an interorganizational domain. *Organization Science, 9*, 217-230.

Harrigan, K. R. (1988). Joint ventures and competitive strategy. *Strategic Management Journal, 9*, 141-158.

Heckman, J. J. (1979). Sample selection bias as a specification error. *Econometrica, 47*(1), 153-161.

Hill, R. C., & Hellriegel, D. (1994). Critical contingencies in joint venture management □ Some lessons from managers. *Organization Science, 5*, 594-607.

Kang, N.-H., & Sakai, K. (2000). *International strategic alliances: Their role in industrial globalisation*. OECD Science, Technology and Industry Working Papers, 2000/05. OECD Publishing. http://dx.doi.org/10.1787/613723204010.

Katila, R. (2002). New product search over time: Past ideas in their prime? *Academy of Management Journal, 45*, 995-1010.

Khanna, T., Gulati, R., & Nohria, N. (1998). The dynamics of learning alliances: Competition, cooperation, and relative scope. *Strategic Management Journal, 19*, 193-210.

Kogut, B. (1988). Joint ventures: Theoretical and empirical perspectives. *Strategic Management Journal, 9*, 319-332.

Li, H., & Atuahene-Gima, K. (2001). Product innovation strategy and the performance of new technology ventures in China. *Academy of Management Journal, 44*, 1123-1134.

Link, A. N. (2002). Private-sector and public-sector strategies to encourage technological alliances. In J. de la Mothe & A. N. Link (Eds.), *Networks, alliances and partnerships in the innovation process* (pp. 7-28). New York, NY: Springer.

Long, J. S., & Freese, J. (2006). *Regression models for categorical dependent variables using Stata*: College Station, TX: STATA Press Publication.

Lyles, M. A., & Salk, J. E. (1996). Knowledge acquisition from foreign parents in international joint ventures: An empirical examination in the Hungarian context. *Journal of International Business Studies, 27*, 877-903.

Medcof, J. W. (1997). Why too many alliances end in divorce. *Long Range Planning, 30*, 718-732.

National Science Board. (2002). *US and international research and development: Funds and alliance □ Science and engineering indicators 2002*. Arlington, VA: National Science Foundation, Division of Science Resources Statistics.

O'Brien, P., Wakeham, W. A., & Walsh, J. T. (1999). University-industry strategic alliances: A British perspective. In Chemical Sciences Roundtable and National Research Council, *Research teams and partnerships: Trends in the chemical sciences: Report of a workshop* (pp. 28-37). Washington, DC: National Academy Press.

OECD. (2004). *Public-private partnerships for research and innovation: An evaluation of the Dutch experience*. Paris, France: Organization for Economic Co-operation and Development.

Olk, P. (1997). The effect of partner differences on the performance of R&D consortia. In P. W. Beamish & J. P. Killing (Eds.), *Cooperative strategies: Vol. 1. American perspectives* (pp. 133-159). San Francisco, CA: New Lexington Press.

Pangarkar, N., & Choo, A. (2001). Do firms seek symmetric alliance partners? An exploratory study. *Journal of Asia-Pacific Business, 3*(1), 83-100.

Pangarkar, N., & Klein, S. (2001). The impacts of alliance purpose and partner similarity on alliance governance. *British Journal of Management, 12*, 341-353.

Pangarkar, N., & Klein, S. (2004). The impact of control on international joint venture performance: A contingency approach. *Journal of International Marketing, 12*(3), 86-107.

Park, S. H., & Russo, M. V. (1996). When competition eclipses cooperation: An event history analysis of joint venture failure. *Management Science, 42*, 875-890.

Park, S. H., & Ungson, G. R. (1997). The effect of national culture, organizational complementarity, and economic motivation on joint venture dissolution. *Academy of Management Journal, 40*, 279-307.

Pearce, R. J. (1997). Toward understanding joint venture performance and survival: A bargaining and influence approach to transaction cost theory. *Academy of Management Review, 22*, 203-225.

Powell, W. W., Koput, K. W., & Smith-Doerr, L. (1996). Interorganizational collaboration and the locus of innovation: Networks of learning in biotechnology. *Administrative Science Quarterly, 41*, 116-145.

Rausser, G., Simon, L., & Ameden, H. (2000). Public–private alliances in biotechnology: Can they narrow the knowledge gaps between rich and poor? *Food Policy, 25*, 499-513.

Roberts, P. W. (1999). Product innovation, product-market competition and persistent profitability in the US pharmaceutical industry. *Strategic Management Journal, 20*, 655-670.

Rothaermel, F. T., & Deeds, D. L. (2006). Alliance type, alliance experience and alliance management capability in high-technology ventures. *Journal of Business Venturing, 21*, 429-460.

Shaver, J. M. (1998). Accounting for endogeneity when assessing strategy performance: Does entry mode choice affect FDI survival? *Management Science, 44*, 571-585.

Sirilli, G., & Evangelista, R. (1998). Technological innovation in services and manufacturing: Results from Italian surveys. *Research Policy, 27*, 881-899.

Stopford, J. M., & Wells, L. T. (1972). *Managing multinational enterprise: Organization of the firm and ownership of the subsidiaries*. New York, NY: Basic Books.

Thomas, J. B., & Trevino, L. K. (1993). Information processing in strategic alliance building: A multiple-case approach. *Journal of Management Studies, 30*, 779-814.

Thursby, J. G., & Thursby, M. C. (2002). Who is selling the ivory tower? Sources of growth in university licensing. *Management Science, 48*, 90-104.

Wernerfelt, B. (1984). A resource-based view of the firm. *Strategic Management Journal, 5*, 171-180.

Whetten, D. A. (1981). Interorganizational relations: A review of the field. *Journal of Higher Education, 52*(1), 1-28.

Wooldridge, J. M. (2009). *Introductory econometrics: A modern approach, 2e*. Mason, OH: South-Western Thomson.

World Bank. (2003). *Productivity and investment climate survey (PICS): Implementation manual.* Working Paper No. 44327, The Private Sector Investment Climate Unit of the World Bank. Retrieved from http://go.worldbank.org/H0DVV557O1

Wu, J. (2011). The asymmetric roles of business ties and political ties in product innovation. *Journal of Business Research, 64*, 1151-1156.

Wu, J. (2012). Technological collaboration in product innovation: The role of market competition and sectoral technological intensity. *Research Policy, 41*, 489-496.

Wu, J. (2014). The effects of external knowledge search and CEO tenure on product innovation: Evidence from Chinese firms. *Industrial and Corporate Change, 23*, 65-89.

Wu, J., & Chen, X. (2012). Leaders' social ties, knowledge acquisition capability and firm competitive advantage. *Asia Pacific Journal of Management, 29*, 331-350.

Zaheer, A., McEvily, B., & Perrone, V. (1998). Does trust matter? Exploring the effects of interorganizational and interpersonal trust on performance. *Organization Science, 9*, 141-159.

ABOUT THE CONTRIBUTORS

Alexander Alexiev is assistant professor of strategic management and innovation at VU University Amsterdam. He specializes in service management, exploratory innovation, and top management teams. He obtained his PhD in strategic management from the Rotterdam School of Management, Erasmus University. He has published in the *Journal of Management Studies* (2010) and the edited volume *The Nature of the New Firm* (Edward Elgar, 2011). E-mail: a.s.alexiev@vu.nl

Marc Bahlmann is assistant professor Faculty of Economics and Business Administration, VU University Amsterdam, Netherlands. He received his PhD from the Faculty of Social Sciences (VU University Amsterdam) for a study on knowledge dynamics among IT entrepreneurs located in the Amsterdam-based IT and new media cluster. His current research interests include alliance portfolios, innovation, and regional innovation. E-mail: m.d.bahlmann@vu.nl

Nora Balogh is strategy consultant at LM Wind Power, Amsterdam, Netherlands. She was awarded Master of Science degrees in economics and business by the University of Groningen and the Corvinus University of Budapest. In her studies, she has specialized in the management of strategic alliances. E-mail: balogh@ymail.com

Ricarda B. Bouncken is a chair professor of strategic management and organization at the University of Bayreuth, Germany, since 2009. Earlier, she was a full professor at the University of Greifswald and the Technical

University of Cottbus. She received her PhD from the University of St. Gall, Switzerland, in 1997. Her research focuses on boundaries across firms and individuals and their effects on innovation. Within the focus of boundaries across firms, she researches vertical and horizontal collaboration and possibilities to improve innovation performance. Within the focus of boundaries across individuals she studies how cultural diversity and age diversity influences work in teams and innovation performance. She has published more than 160 articles and books and leads several large research grant projects. E-mail: bouncken@uni-bayreuth.de

Leonardo Corbo is a post-doc research fellow in strategic management at LUISS Guido Carli University, Italy. He obtained his PhD in Business Administration from the University of Bologna, Italy. He has been a visiting scholar at Kellogg School of Management, Northwestern University. His research interests include alliances and networks and organizational responses to unexpected events. One of his papers was shortlisted among the final ten papers for the Best PhD Paper Prize at the 2011 Strategic Management Society Conference. Prior to joining academia, he worked in the marketing department at FIAT's headquarters in Turin, Italy. E-mail: lcorbo@luiss.it

T. K. Das is professor of strategic management at the Zicklin School of Business, Baruch College, City University of New York. He is concurrently a member of the university's doctoral faculty. Professor Das received his PhD in organization and strategic studies from the Anderson Graduate School of Management, University of California at Los Angeles (UCLA). He also has degrees in physics, mathematics, and management, and a professional certification in banking. Prior to entering the academic life, Professor Das had extensive experience as a senior business executive. He has research interests in strategic alliances, strategy making, organizational studies, temporal studies, and executive development. Professor Das has published over a dozen books and monographs, and his research has appeared in over 45 journals, of which some of the later ones include *Academy of Management Executive*, *Academy of Management Review*, *British Journal of Management*, *Journal of International Management*, *Journal of Management*, *Journal of Management Studies*, *Organization Science*, *Organization Studies*, and *Strategic Management Journal*. Professor Das currently serves on the editorial boards of several academic journals, and is a former senior editor of *Organization Studies* and board member of a number of other journals. He is the founding (and current) series editor of the two book series, *Research in Strategic Alliances* and *Research in Behavioral Strategy* (Information Age Publishing). E-mail: TK.Das@baruch.cuny.edu

Gjalt de Jong is an associate professor of strategy at the Faculty of Economics and Business, University of Groningen, Netherlands. His research and publication relate to the management of strategic alliances; business ethics in transition economies, the structure and organization of multinational enterprises, and the causes and consequences of national rules. He is responsible for the design and coordination of courses in the fields of International business, international economics, and public administration. Prior to his current position, he was affiliated with PricewaterhouseCoopers and KPMG as a senior management consultant, and was responsible for large-scale strategy programs at Dutch multinational firms in industry and the financial sector, as well as for educational institutes. E-mail: g.de.jong@rug.nl

Ard-Pieter de Man is professor of knowledge networks and innovation at VU University Amsterdam and dean of SIOO. His main research area is in alliances and networks. He is a member of the Global Board of the Association of Strategic Alliance Professionals and has consulted with numerous organizations on alliance management. E-mail: a.p.deman@vu.nl

Geert Duysters is a full professor of entrepreneurship and innovation at Tilburg University. He currently acts as the scientific director of the Tilburg Center of Entrepreneurship. His academic research mainly concerns corporate entrepreneurship, international business strategies, innovation strategies, mergers and acquisitions, network analytical methods and strategic alliances. He has published over 100 international refereed articles and book chapters in among others: *Organization Science, Journal of International Business Studies, Research Policy, Organization Studies, Journal of Management Studies.* E-mail: gduysters@tilburguniversity.edu

Isabel Estrada is assistant professor in the Department of Innovation Management and Strategy at the University of Groningen, Netherlands, where she currently teaches in the area of organizational innovation. Earlier, she enjoyed a Contract for Novel Researchers (2008-2012) at the University of Valladolid, Spain, co-funded by the "Junta de Castilla y León" and the UE Social Fund. She received her PhD in 2012 from the University of Valladolid, Spain, with a dissertation titled "Collaboration and value creation in multipartner R&D alliances: A longitudinal case study on the Acuisost Consortium." Her research focuses on the processes of exchange in interorganizational collaboration, with a particular interest in multipartner settings. E-mail: I.Estrada.Vaquero@rug.nl

Anne Haugen Gausdal is an associate professor at Vestfold University College, Department of Maritime Technology and Innovation, Norway. She received her PhD from Bodø Graduate School of Business, Norway, in 2008. Her publications relate to network development, innovation, trust, and methods for knowledge-sharing and trust-building. In the autumn of 2012 she was a visiting scholar at Jacobs University, Bremen, Germany. In the spring of 2013 she was a visiting scholar at Eastern Finland University, Joensuu, Finland. She is an editorial board member of the *International Entrepreneurship and Management Journal* and frequently reviews for four other international journals. Gausdal is currently the manager of the VRI Vestfold and Buskerud research project for Regional R&D and Innovation, Norway. Her field of interest includes networks, strategic alliances, trust and innovation. E-mail: ang@hive.no

Mark Greeven is associate professor in the Management Science and Engineering Department of the School of Management at Zhejiang University, China. His PhD is from Erasmus University, Netherlands. He is a research fellow at the National Institute for Innovation Management, China, and the International Institute for Asian Studies (Centre for Regulation and Governance), Netherlands. He is a member of the editorial board of "China in transition," a book series of Parthenon. Before joining ZJU-SOM, he was assistant professor at the Rotterdam School of Management and co-founder of Rotterdam School of Management's (RSM) master's program "Chinese Economy and Business" (in cooperation with Leiden University, Zhejiang University, and Lingnan (University) College and over 30 enterprises). His research interests include innovation, entrepreneurship and foreign R&D in emerging economies. He also explores the strategies of Chinese firms going abroad and cooperation with foreign firms in a joint project with RSM (Holland) and China Europe International Business School. He has published in journals such as *Technology Analysis & Strategic Management*, *European Management Journal*, and *International Journal of Services Technology and Management*, and has also authored book chapters and written for newspapers such as *Financial Times*. E-mail: mgreeven@zju.edu.cn

Michael Hitt is a distinguished professor of management at Texas A&M University and holds the Joe B. Foster Chair in Business Leadership. He received his PhD from the University of Colorado. Dr. Hitt has coauthored or co-edited 26 books and authored or coauthored many journal articles. A recent article in the *Journal of Management* listed him as one of the 10 most cited authors in management over a 25-year period. Additionally, the *Times Higher Education* in 2010 listed him among the top scholars in economics, finance and management based on the number of articles with a high citation rate on the Web of Science. He has served on the editorial review boards of multiple journals and is a former editor of the *Academy of*

Management Journal and the *Strategic Entrepreneurship Journal*. He is a fellow in the Academy of Management and in the Strategic Management Society, a research fellow in the National Entrepreneurship Consortium and received an honorary doctorate from the Universidad Carlos III de Madrid. He is a former president of the Academy of Management, a former president of the Strategic Management Society and a member of the Academy of Management Journals' Hall of Fame. He received awards for the best article published in the *Academy of Management Executive* (1999), *Academy of Management Journal* (2000), and the *Journal of Management* (2006). He received the Irwin Outstanding Educator Award and the Distinguished Service Award from the Academy of Management and the Falcone Distinguished Entrepreneurship Scholar Award from Syracuse University. E-mail: mhitt@mays.tamu.edu

Jun Huang is professor in the College of Economics and Management at Southwest University, China. He has published over 20 articles in various journals, including *British Journal of Management, Journal of Global Marketing,* and *International Journal of Human Resource Management.* His research interests include strategic management and the Chinese economy. E-mail: john0227@hotmail.com

Paul Iske is chief dialogues officer at ABN AMRO Bank and professor of open innovation and business venturing at Maastricht University, Netherlands. His research interests include venturing, knowledge management, service innovation, and co-creation. Furthermore, he is a boardroom consultant and speaker in these areas. E-mail: p.iske@maastrichtuniversity.nl

Rosalinde Klein Woolthuis is associate professor at the Faculty of Economics and Business Administration at VU University Amsterdam, Netherlands, and is a consultant at TNO Innovation and Environment, Delft, Netherlands. She has published papers in journals such as *Organization Studies*, *Technovation*, and *Technological Forecasting & Social Change*. Her research focuses on the diffusion of sustainable innovations and practices, thereby changing existing business models and entire industries. She is a consultant to national and international governmental bodies on how sustainable innovation can be promoted through measures such as stimulating entrepreneurship and modernizing rules and regulations. E-mail: rkleinwoolthuis@feweb.vu.nl

Peter Ping Li is professor of Chinese Business Studies at Copenhagen Business School, Denmark. Earlier, he was Professor of Management at California State University, Stanislaus, in the United States. He has published over 30 articles in various academic journals, including *Organization*

Studies, Journal of Operations Management, Management and Organization Review, Journal of International Management, Asia Pacific Journal of Management, and *Journal of Business Ethics.* He has been widely recognized as one of the leading scholars in two fast-emerging research streams: (1) emerging multinational firms (including the subareas of transaction value theory, learning-based view of MNE, and organizational trust), and (2) indigenous research on Chinese management (including the sub-areas of "yin-yang balancing" and "wu" as intuitive imagination as well as their applications to all management paradoxes, tensions, and dilemmas). E-mail: ppl.int@cbs.dk

Natalia Martín-Cruz is associate professor at the Faculty of Business and Economics in the University of Valladolid, Spain, where she teaches strategic management. Currently holding the chair of the Department of Management, she became doctor in business economics (2000, University of Valladolid). In her dissertation, she compared the institutional frameworks of the United States and Spanish pharmaceutical industries. She has conducted research in a variety of topics (e.g., entrepreneurship, teamwork, knowledge sharing, corporate spin-offs, and strategic alliances), visiting several external institutions, such as the Institute of Business Innovation (University of California at Berkeley) and Stanford Institute for Economic Policy Research (Stanford University). E-mail: ambiela@eco.uva.es

Victor M. Martín-Pérez is associate professor at the Faculty of Business and Economics in the University of Valladolid, Spain, where he teaches courses of Operations Management and Strategy. An expert in the organizational architecture of non-profit organizations, he earned his PhD in Business Economics (University of Valladolid, 2006), subsequently visiting the University of Texas-Pan American (Edinburg) for postdoctoral research. Aside from studying the impact of business simulation games in students' learning skills, he has mainly investigated how the processes of delegation and motivation affect knowledge transfer and efficiency in nonprofit organizations. E-mail: vmartin@eco.uva.es

Paul Olk is a professor of management and the assistant dean of faculty, research and accreditation at the Daniels College of Business of the University of Denver, Denver. His research on strategic alliances has appeared in numerous journals and books. He is a co-author of *Entrepreneurial Alliances* (Pearson Publishing), the editor of *Strategy Process Research* (Edward Elgar Publishing), and serves or has served on several editorial review boards. He has served as the chair of the Technology and Innovation Management Division of the Academy of Management and is a past organizer of two *Organization Science* Winter Conferences. He received his PhD from The Wharton School of Business at the University of Pennsylvania. E-mail: Paul.Olk@du.edu

Nina Veflen Olsen is full time senior research scientist at Nofima, the Norwegian Institute of Food, Fishery and Aquaculture, and part time researcher at VRI Buskerud, the Program for Regional R&D and Innovation, Norway. She received her PhD from BI—The Norwegian Business School in 2006, has published extensively within the area of consumer driven food product development and is currently head of the innovation research group at Nofima. She is an experienced research manager for large national and international research projects. Among other things she was work package leader within EUs sixth framework for QPorkChains (improving the quality of pork and pork products for the consumer), NovelQ (novel processing methods for the production and distribution of high-quality and safe food) and DoubleFresh (towards a new generation of healthier and tastier ready meals with fresh ingredients), and frequently serves as a reviewer for journals such as *Appetite, Food Quality and Preference*, and *British Food Journal*. Her fields of interest include strategic alliances in new product development and open innovation. E-mail: nina.veflen.olsen@nofima.no

Nitin Pangarkar received his PhD from the University of Michigan, Ann Arbor (United States) in 1993. He teaches strategy and international business courses in the BBA, MBA and executive MBA programs at the National University of Singapore, where he has received numerous awards for excellence in teaching and service. His research interests lie in the areas of strategic management and international business, specifically, cross-border strategic alliances, mode of entry and global strategy. His research has been published in journals such as *Strategic Management Journal, Journal of Management Studies, Journal of World Business, Journal of International Marketing, International Business Review, Long Range Planning*, and *Asia Pacific Journal of Management*. His research has received awards in academic conferences as well as for its practical implications. He is a sole or co-author of two managerial books and one casebook. He has also authored numerous columns and commentaries on current business events in the media in Singapore, Malaysia, India, and Thailand, among others. E-mail: bizpn@nus.edu.sg

Robin Pesch is a PhD student and research assistant to the chair of Strategic Management and Organization of the University of Bayreuth, Germany, since November 2009. He researches the management of vertical innovation alliances and presented his findings on several international conferences (e.g., AOM, AIB, WAMS). For his PhD thesis he is researching the effects of cultural distance on different forms of interorganizational learning in international innovation alliances. His dissertation proposal reached the final round of the Dissertation Proposal Competition 2012 of the Product Development and Management Association (PDMA). E-mail: robin.pesch@uni-bayreuth.de

Jaspar Roos is chief inspiration officer for Dialogues Incubator, the ABN AMRO Bank corporate venturing accelerator for open innovation in the financial industry. He is also part of numerous advisory and executive boards, such as the Erasmus Medical Center Incubator, FEI EMEA, and TEDxBaghdad. He also runs a public-private partnership called Future-Ideas.eu. E-mail: jaspar.roos@dialoguesincubator.nl

Weilei (Stone) Shi is an assistant professor of strategy at Zicklin School of Business, Baruch College, City University of New York. He obtained his PhD in strategic management from the University of Pittsburgh. His research centers on the interaction between strategy and international management. Specifically, he is interested in examining M&As and alliances from both temporal and network perspectives. His papers have been published in *Strategic Management Journal*, *Journal of International Business Studies*, *Journal of Management*, *Journal of Management Studies*, *Organization Studies*, *Academy of Management Perspectives*, and the *Academy of Management Best Paper Proceedings*. He also has chapters in two book series, *Advances in Mergers and Acquisitions* and *Research in Strategic Alliances*. He received an Outstanding Reviewer Award in the annual Academy of Management conferences from 2005 to 2007. Before joining academia Stone worked for Roland Berger Strategy Consultancy, a global strategy consulting firm in Shanghai, China. E-mail: weilei.shi@baruch.cuny.edu

Regien Sumo is a PhD candidate at Eindhoven University of Technology, Netherlands. She obtained a master's degree in strategic management from Tilburg University, Netherlands, writing her thesis on corporate venture capital and real options valuation. During her graduate studies, she studied financial investments at the University of California at Berkeley, United States. Her current research interests include formal governance and innovation in interorganizational relationships. E-mail: R.A.F.Sumo@tue.nl

Jing Sun is a PhD candidate in strategic management at Joseph M. Katz Graduate School of Business, University of Pittsburgh. Her research interests include strategic usage of social media, strategic alliance networks, mergers and acquisitions and temporal perspective. Her work has recently appeared in the *Journal of Management*. Prior to joining the PhD program, she has worked in managerial positions for Mercedes Benz Financial Services and Beijing Benz Automotive Co., both in Beijing, China. E-mail: jingsun@katz.pitt.edu

Brian Tjemkes is an associate professor at the Faculty of Economics and Business Administration, at the VU University Amsterdam, Netherlands. He has published in *Journal of Management Studies*, *Management Decision*, and *Journal of International Management*, among others. He is also co-author of the book *Strategic Alliance Management*, published by Routledge (2012). His research interests include the management and performance of strategic alliances. E-mail: b.v.tjemkes@vu.nl

Geerten van de Kaa is assistant professor of strategy and innovation at Delft University of Technology, Netherlands. He holds a PhD from Rotterdam School of Management, Erasmus University. His research interests include platform wars for complex systems, smart and networked governance, and responsible innovation. In 2011, he was a visiting scholar at Berlin University of Technology. He has published in high ranking international journals, including *Organization Studies*, *IEEE Transactions on Engineering Management*, *Technological Forecasting and Social Change*, *Technology Analysis & Strategic Management*, *Renewable and Sustainable Energy Reviews* and *Computer Standards & Interfaces*. In 2009, he won the best paper award at the Standardization and Innovation in Information Technology conference and in 2010 he won the NNK best doctoral dissertation award. He teaches courses in the strategic management of technological innovation.

Wendy van der Valk is an assistant professor of supply chain management at the Tilburg School of Economics and Management, Tilburg University, Netherlands. She obtained her PhD from the Rotterdam School of Management, Erasmus University, Netherlands, where she conducted her doctoral research on buying/outsourcing business services. Her research focuses on the purchase of business-to-business services, specifically in the context of triadic constellations between buyers, suppliers and customers. Among her current research topics is the use and effects of performance-based contracts in service triads. Her work has been published in, among others, *International Journal of Operations & Production Management*, *Journal of Supply Chain Management*, and *Industrial Marketing Management*. E-mail: W.vdrValk@uvt.nl

Arjan van Weele holds the NEVI (Dutch Association for Purchasing Professionals) chair of purchasing and supply management at Eindhoven University of Technology, Netherlands. He worked for many years in education and research programs at Eindhoven University of Technology and Nyenrode University. He is/was visiting professor in several management development programs, in such instituions as TIAS-Nimbas School of Management, Netherlands, Vlerick Gent Leuven School of Management, Belgium, University of Stuttgart, Germany, European Institute of Purchasing Management, France, Kellogg School of Business at Northewestern University, United States, and IMD (International Institute for Management Development), Switzerland. Based upon his extensive research and publications he qualified as a senior research fellow of BETA, the accredited research school of Eindhoven University of Technology and Twente University, Netherlands. He is an editorial board member of *Journal of Purchasing and Supply Management* and a member of the advisory board of the *Journal of Supply Chain Management*. He holds an MBA degree from Interfaculteit Bedrijfskunde Delft, Netherlands, and a PhD from Eindhoven University of Technology. He has published over 15 books and 150 articles on purchasing and supply management in refereed journals such as *Journal of Purchasing and Supply Management, Journal of Supply Chain Management, R&D Management, Journal of Product Innovation Management, IEEE Transactions on Engineering Management, Industrial Marketing Management, Supply Chain Management: An International Journal*, and *Journal of International Management*. In 2003 he received the Hans Ovelgönne Award from the International Federation of Purchasing and Supply Management (IFPSM) for his outstanding academic contribution in the area of purchasing and supply management. E-mail: A.J.v.Weele@tue.nl

Jie Wu received his PhD in business policy from National University of Singapore. He is an assistant professor of management and marketing at the University of Macau, China. His research has been published in journals such as *Research Policy, Industrial & Corporate Change, International Journal of Research in Marketing, Management and International Review, Journal of Business Research, Industrial Marketing Management, International Business Review, Long Range Planning, Asia Pacific Journal of Management,* and *Management Decision*. He has also authored four book chapters. He has been a recipient of the best doctoral paper award of the 2008 Asia Academy of Management conference in Taipei and was a finalist for the Haynes Prize of the 2011 Academy of International Business in Nagoya. His papers have been selected for the Best Papers Proceedings of the 2009 and 2012 Academy of Management annual meetings. His current research focuses on the internationalization and growth of emerging market firms, strategic alliances, and innovation. E-mail: Jiewu@umac.mo

Jie Xue is an associate professor of management at the Department of Management Science at South China Normal University, China. His current research involves regional innovation systems, knowledge networks, and high technology clusters in Pearl River Delta in south China. He received his PhD in management science and engineering from the School of Business Administration at South China University of Technology, China. E-mail: xuejiesky@gmail.com

Jie Zhou is associate professor in the School of Management, Southwest University of Political Science and Law, China. He has published 20 articles in various academic journals in China. His academic interest lies in strategic alliances, knowledge management, and innovation. E-mail: jiezhou929@163.com

INDEX

A

B

D

E

F

G

H

I

J

K

L

M

N

Q

R

S

T

U

V

W

X

Y

Z

www.ingramcontent.com/pod-product-compliance
Lightning Source LLC
LaVergne TN
LVHW052253200926
848481LV00002B/359
* 9 7 8 1 6 2 3 9 6 6 2 2 5 *